T0296429

Handbook of Infrared Astronomy

Infrared astronomy is a dynamic area of current research. It has been revolutionized in the past few years by the advent of large, sensitive, infrared arrays and the success of several infrared satellites. This handbook provides a clear, concise and accessible reference on all aspects of infrared astronomy. Throughout the book, the emphasis is on fundamental concepts, practical considerations and useful data.

Starting with a review of the basic infrared emission mechanisms, we are shown how the Earth's atmosphere affects and limits observations from ground-based telescopes. The important systematics of photometric accuracy are treated in detail. Spectroscopy – both stellar and otherwise – is explained, and illustrated with useful examples. An important chapter is devoted to dust, which plays such a central role. Finally, the technical background to infrared instrumentation is covered to help the reader develop a proper understanding of the capabilities and limitations of infrared observations.

This volume provides both an essential introduction for graduate students making infrared observations or reducing infrared data for the first time, and a convenient reference for more experienced researchers.

Ian Glass obtained his first degree at Trinity College Dublin and his PhD at the Massachusetts Institute of Technology. Following postdocs at MIT and Caltech, he spent five years at the Royal Greenwich Observatory. He has authored over 170 papers in astronomical journals and conference proceedings and is currently Head of Instrumentation at the South African Astronomical Observatory. Also fascinated by astronomical history, he has recently written *Victorian Telescope Makers: the Lives and Letters of Thomas and Howard Grubb* (IOP Publishing, Bristol and Philadelphia, 1997).

Cambridge Observing Handbooks for Research Astronomers

Today's professional astronomers must be able to adapt to use telescopes and interpret data at all wavelengths. This series is designed to provide you with a series of concise, self-contained handbooks which cover the basic principles peculiar to observing in a particular spectral region, or to using a special technique or type of instrument. The books can be used as an introduction to the subject, and as a handy reference – for use at the telescope, or in the office. They also promote an understanding of other disciplines in astronomy and a modern, multi-wavelength, multi-technique approach to research. Although aimed primarily at graduate students and researchers, many titles in the series are of interest to keen amateurs and undergraduate students.

Series editors
Professor Richard Ellis, Institute of Astronomy, *University of Cambridge*
Professor John Huchra, Center for Astrophysics, *Smithsonian Astrophysical Observatory*
Professor Steve Kahn, Department of Physics, *Columbia University*, New York
Professor George Rieke, Steward Observatory, *University of Arizona*, Tucson
Dr. Peter B. Stetson Herzberg, Institute of Astrophysics, *Dominion Astrophysical Observatory*, Victoria, British Columbia

HANDBOOK OF INFRARED
ASTRONOMY

I. S. GLASS
SOUTH AFRICAN ASTRONOMICAL OBSERVATORY

CAMBRIDGE
UNIVERSITY PRESS

PUBLISHED BY THE PRESS SYNDICATE OF THE UNIVERSITY OF CAMBRIDGE
The Pitt Building, Trumpington Street, Cambridge, United Kingdom

CAMBRIDGE UNIVERSITY PRESS
The Edinburgh Building, Cambridge CB2 2RU, UK
40 West 20th Street, New York, NY 10011-4211, USA
10 Stamford Road, Oakleigh, Melbourne 3166, Australia

www.cambridge.org
Information on this title: www.cambridge.org/9780521633116

First published 1999

Typeset in Computer Modern in LATEX [TB]

*A catalog record for this book is available from
the British Library*

Library of Congress Cataloging-in-Publication Data
Glass, I. S. (Ian S.)
    Handbook of infrared astronomy / I.S. Glass.
        p.   cm. – (Cambridge observing handbooks for research
    astronomers)
      ISBN 0-521-63311-7
      1. Infrared astronomy.   I. Title.   II. Series.
    QB47.G53 1999
    522′.683–dc21   98-4780
                                                              CIP

ISBN-13 978-0-521-63311-6 hardback
ISBN-10 0-521-63311-7 hardback

ISBN-13 978-0-521-63385-7 paperback
ISBN-10 0-521-63385-0 paperback

Transferred to digital printing 2005

To Hettie

Celestial rosy red, love's proper hue

Contents

Introduction

More and more investigators are being attracted to work in the infrared spectral region. My intention in this book is to introduce the infrared, its peculiarities and its special techniques, to a new audience that may include established astronomers as well as new students. Basic facts and figures are emphasized and I have not tried to cover all current research. The approach usually avoids the historical and the reader is referred instead to recent books and papers. The origins of ideas are always complex and, with something like 10^4 papers published, it is impossible to do justice to every individual in the field.

The first chapter deals with the basic facts of blackbody radiation, which plays a dominant role in infrared astronomy. A short review of atomic and molecular physics follows in order to provide a convenient reminder of the meaning of spectroscopic nomenclature. The second chapter covers the general properties of the Earth's atmosphere and outlines the main infrared surveys that have taken place or will shortly do so. Chapter 3 is devoted to photometry, emphasizing its fundamental aspects and the importance of traceability and calibration. Chapter 4 is an introduction to spectroscopy, treating firstly the stars. Photodissociation regions, which are of great interest in the infrared and millimeter region, are followed by HII regions and some "Rosetta Stone" spectra of representative objects. Chapter 5 is devoted to dust and its central role in star formation. Finally, because an understanding of the technology is important in obtaining reliable results, the last chapter covers the basics of infrared instrumentation.

Each author of a book in the COHRA series has been asked to name some of the milestones that, in his or her view, line the road of discovery in his subject. My personal choice is based on the feeling that progress in infrared astronomy has been directly attributable to technological advances and new types of instruments.

1

Milestones of infrared astronomy

- Discovery of infrared radiation whilst examining the warming powers of the Sun's rays as dispersed by a prism (Sir William Herschel, 1800).
- The use of PbS detectors, a product of research during World War II, and newly developed interference filters, by H. L. Johnson and collaborators to set up the first useful photometric system (*ca* 1962).
- The development of the gallium-doped germanium bolometer and all-metal experimental dewars by F. J. Low (*ca* 1961), as well as their application to astronomical photometry soon afterwards.
- The Two-Micron Sky Survey by G. Neugebauer and R. B. Leighton which revealed the existence of stellar objects that radiated predominantly in the infrared (*ca* 1965).
- The launch of the IRAS cryogenically cooled infrared satellite (1983), which permitted the first reliable long-wavelength survey unlimited by the Earth's atmosphere.
- The launch of the ISO pointable observatory-style satellite with spectroscopic instrumentation (1995) that has allowed the detailed examination of specific objects.

Present-day progress

At the present moment, new near-infrared surveys reaching to faint magnitudes, such as DENIS and 2MASS, are revealing a wealth of celestial detail, while IRAS and ISO have opened the mid- and far-infrared wavelength regions to systematic investigation and serendipitous discovery, leading to follow-up studies through the whole observable spectrum.

Infrared arrays have increased in size and sophistication to the point where they offer resolutions and quantum efficiencies similar to the best available in the visible spectrum. The new instruments being designed to exploit them will enable the study of objects in great spatial and spectroscopic detail.

New ground-based telescopes with very large collecting areas are being completed at favourable high-altitude sites, leading to hitherto unattainable sensitivities. Advances in active and adaptive optics have opened new possibilities in the studies involving small angular size. The infrared offers particular advantages for this type of work.

The future

The next few years will see many observations which will exploit the obvious advantages of the new instrumentation. A few of the areas awaiting investigation are:

- The study of primeval galaxies and QSOs at high redshifts.
- The life-cycle of dust in the interstellar medium of galaxies and its relation to the presence of heavy elements at early cosmic times.
- The determination of the shape of the Milky Way galaxy, which has remained a difficult problem because of obscuration by interstellar dust in the Galactic Plane.
- The better understanding of stellar distributions and physical processes occurring near the Galactic Centre, also inhibited hitherto by obscuration.
- The study of the boundary zones between HII regions and molecular clouds – the photodissociation regions.
- The discovery of brown dwarfs and planetary candidates, which are too cool to radiate significantly at visible wavelengths.
- Application of infrared photometry to distance indicators: the often unpredictable systematic effects of interstellar reddening being minimal in the infrared.

Acknowledgments

I would like to thank the following people for their suggestions and comments, while not holding them responsible for any errors that may remain in the text!

Prof. M. S. Bessell, Mount Stromlo Observatory
Prof. M. W. Feast, University of Cape Town
Dr. W. J. Forrest, University of Rochester
Dr. C. D. Laney, South African Astronomical Observatory
Dr. T. Lloyd Evans, South African Astronomical Observatory
Mr. Paul Martini, Ohio State University
Dr. P. J. McGregor, Mount Stromlo Observatory
Dr. A .F. M. Moorwood, European Southern Observatory
Prof. G. H. Rieke, University of Arizona
Prof. D. C. B. Whittet, Rensselaer Polytechnic Institute

The following are thanked for permission to reproduce figures and tables:

Academic Press: 2.3; Dr. P. André, CEA Saclay: 5.4; Astronomy and Astrophysics: 2.7, 3.8, 4.2, 4.3; Prof. M. Barlow, Univ. Coll. London: Table 4.9; Prof. M. Bessell, MSSSO: 3.9; Cambridge Univ. Press: 5.1, 6.4; Dr. R. Chini, MPIR: 5.6; Editions Frontières: 4.1, 5.4; Prof. R. Genzel, MPE: 5.5; Drs. S. Lord, IPAC, and A. Tokunaga, Univ. of Hawaii: 3.4; Dr. Dieter Lutz, MPE: Table 4.10; Dr. M. F. Kessler and Mr. Ray Long, ESA: 6.1; Kluwer Academic Publishers: 5.3, 5.6; Monthly Notices of the Royal Astronomical Society: 2.4–2.6; Dr. A. F. M. Moorwood, ESO: 4.1; Publications of the Astronomical Society of the Pacific: 3.1, 6.7; Tables 3.1, 3.2; Raytheon Infrared Center of Excellence: 2.1; Dr. T. Simon, IFA, Hawaii: 3.5; Dr. D. Simons, NOAO: 3.2, 3.3; Dr. I. Yamamura, Univ. of Tokyo: 5.3.

If familiarity has led to a disproportionate number of references to my own work, I hope I will be forgiven. Finally, readers who notice errors or who would like to make suggestions for future editions are invited to contact me at isg@saao.ac.za.

I. S. Glass
Cape Town

1

Infrared Emission Mechanisms

1.1 Some photometric definitions

1.1.1 Source-related terms

1.1.1.1 Luminosity

The total outward flow of radiation from a body is called its *luminosity L*, measured in W. If divided into spectral intervals, this quantity becomes its *monochromatic luminosity L_ν*, measured in units of W Hz^{-1}, or L_λ, measured in units of W μm^{-1}.

1.1.1.2 Intensity

The power emitted from unit area of a source into unit solid angle is called the *intensity I* and it is measured in W sterad^{-1} m^{-2}. The same quantity, when divided into spectral intervals, is called the *specific intensity I_ν*; similarly for I_λ.

1.1.2 Receiver-related terms

1.1.2.1 Flux density

The received radiation, F_ν or F_λ per unit of frequency or wavelength, is measured in terms of W m^{-2} Hz^{-1} or W m^{-2} μm^{-1}.[†] This quantity is known as the *monochromatic flux density* or simply the *monochromatic flux* in astronomical parlance.

The monochromatic flux density is also called the *spectral irradiance*.

Some other photometric terms used outside astronomical circles are given by Sterken and Manfroid (1992) in their general book on photometry.

[†] The MKS unit is W m^{-2} m^{-1}.

Table 1.1. *Some useful physical constants and other quantities*

Symbol	Quantity	Value
c	Velocity of light	2.998×10^8 m s^{-1}
h	Planck's constant	6.626×10^{-34} J s
k	Boltzmann's constant	1.380×10^{-23} J $^\circ$K^{-1}
e	Electronic charge	1.602×10^{-19} Coulomb
$L_\odot$	Solar luminosity	3.83×10^{26} W
$M_\odot$	Solar mass	1.989×10^{30} kg
σ	Stefan's constant	5.669×10^{-8} W m^{-2} $^\circ$K^{-4}
π		3.14159
e	Base of natural logs	2.71828

Flux densities in frequency units can be converted into wavelength units and vice versa. The flux density in unit frequency interval is F_ν. Since

$$\lambda = \frac{c}{\nu}$$

we have

$$\mathrm{d}\lambda = -\frac{c}{\nu^2}\mathrm{d}\nu = -\frac{\lambda^2}{c}\mathrm{d}\nu.$$

If we put $\mathrm{d}\lambda = 1$, $\mathrm{d}\nu$ becomes the frequency interval corresponding to unit wavelength interval and we see that

$$F_\lambda = 2.998 \times 10^8 F_\nu/\lambda^2$$

if λ is expressed in metres and the other quantities are in MKS units.

It is also interesting to note that

$$\nu F_\nu = \lambda F_\lambda$$

if λ is again expressed in metres. In discussing the overall energy distribution of the radiation from an astronomical object, $\log \nu F_\nu$ or $\log \lambda F_\lambda$ is often graphed against $\log \nu$ or $\log \lambda$ to show in what frequency or wavelength regime it emits the most power per decade.

1.1.3 Optical depth

If I is the intensity of a ray of original intensity I_0 which has passed through a layer of absorbing material, then the *optical depth* τ of the material is given by

$$I = I_0 e^{-\tau}.$$

For $\tau = 1$, the ray emerges at $e^{-1} \simeq 0.368$ times its original strength. More specifically, we can write

$$I_\lambda = I_{\lambda_0} e^{-\tau_\lambda}.$$

1.2 Blackbodies

1.2.1 Planck distribution

A *blackbody* is one which absorbs all the radiation which falls on it, i.e., it has *absorptivity* equal to 1. The *emissivity* of a body is the energy it emits per unit time as a fraction of what it would emit if it were a blackbody.

Kirchhoff's law states that the ratio of the emissivity to the absorptivity of a body at a given wavelength depends only on its temperature and not on its nature.

If an object is to radiate like a blackbody, it must be optically thick at the wavelengths concerned. For example, a dwarf star like our Sun has a well-defined thin layer (the photosphere) whose optical depth is large at visible and ultraviolet wavelengths. A Mira variable, on the other hand, has an extended diffuse atmosphere, and the optical depth at a given radius is a strong function of wavelength, especially in the visible region. Measured at the wavelength of a strong absorption line or band, a Mira appears to have a much larger radius than it has in the continuum.

In the laboratory, blackbody radiation is created as electromagnetic radiation in equilibrium with the walls of an enclosure kept at temperature T, for example, the inside of a furnace with constant-temperature walls. Even if the walls are not perfectly absorbing, so that only some fraction of the incident radiation is absorbed by them, the fact that it is enclosed ensures that the radiation density within is very close to that of a blackbody.

In an astrophysical source, the surface facing the observer can often be regarded as approximating a blackbody of a particular temperature. The emitted spectrum deviates more or less from the true blackbody spectrum according to the variation of the optical depth with wavelength and the temperature gradient near the surface.

In the case of the surface of a blackbody at temperature T, the specific intensity per unit frequency interval is given by the fundamental relationship, first derived theoretically by Max Planck (1858–1947) in 1900:

$$B_\nu = \frac{2h}{c^2}\frac{\nu^3}{e^{h\nu/kT}-1}\ \text{W}\,\text{m}^{-2}\,\text{Hz}^{-1}\,\text{sterad}^{-1}.$$

The equivalent in wavelength units is

$$B_\lambda = 10^{-6}\frac{2hc^2}{\lambda^5}\frac{1}{e^{hc/\lambda kT}-1}\ \text{W}\,\text{m}^{-2}\mu\text{m}^{-1}\text{sterad}^{-1},$$

where c = velocity of light, h = Planck's constant and k = Boltzmann's constant. The wavelength λ and the constants are in MKS units.

To obtain the monochromatic luminosity *per unit surface area* of a blackbody, B_ν or B_λ must be multiplied by π.

1.2.2 *Wien displacement law*

The wavelength λ_{peak}, measured in μm, at which the maximum of the blackbody specific intensity, B_λ, occurs, is related to the temperature (in K) by

$$T\lambda_{\text{peak}} = 2898.$$

This relationship is known as the *Wien displacement law*.

Similarly, for frequency units, ν_{max}, the frequency in Hz at which the maximum of B_ν occurs is

$$\nu_{\text{max}} = 5.878 \times 10^{10}T.$$

1.2.3 *Rayleigh–Jeans approximation*

For the case of wavelengths much longer than that at which B_λ is at a maximum,

$$B_\nu = \frac{2kT}{c^2}\nu^2\ \text{W}\,\text{m}^{-2}\,\text{Hz}^{-1}\,\text{sterad}^{-1}$$

or

$$B_\lambda = 10^{-6}\frac{2ckT}{\lambda^4}\ \text{W}\,\text{m}^{-2}\mu\text{m}^{-1}\text{sterad}^{-1};$$

this is known as the *Rayleigh–Jeans approximation*. In radio astronomy, it is customary to speak of the *antenna temperature* due to a source,

based on this law. For a source of uniform surface brightness with angular extent greater than the beamwidth of the antenna, the measured
intensity translates directly into a temperature.

Also arising from radio astronomical usage is the *Jansky* as a unit of
monochromatic flux density. One Jansky is 10^{-26} W m^{-2} Hz^{-1}.

1.2.4 Stefan's law

The luminosity per unit area of a blackbody is

$$\frac{\sigma T^4}{\pi}$$

where σ is known as Stefan's constant and has the value 5.669 $\times$ 10^{-8}
W m^{-2} K^{-4}.

Stefan's constant is obtained by integrating the blackbody distribution function over all frequencies and is given in terms of fundamental
physical constants as follows:

$$\sigma = \frac{2\pi^5 k^4}{15c^2 h^3}.$$

1.2.5 Radiation density

The energy density u_ν of radiation, or *radiation density*, in a region
surrounded by a blackbody of uniform temperature is given by:

$$u_\nu = \frac{4\pi B_\nu}{c} \text{ J m}^{-3} \text{ Hz}^{-1}$$

and, when integrated over frequency,

$$U = \int u_\nu \, d\nu = \frac{4\sigma}{c} T^4 \text{ J m}^{-3},$$

where σ is Stefan's constant.

1.2.6 Infrared colours of blackbodies

Many objects radiate approximately like blackbodies. Some stars have
photospheres which approximate a blackbody of a certain temperature
and are surrounded by thin dust shells which radiate like blackbodies of
a lower temperature. The accompanying Table 1.2 lists the broad-band
infrared colors of some representative blackbodies.

Table 1.2. *Broad-band colors of blackbodies*

Temp (K)	J-H	H-K	K-L	L-N
300	9.37	6.92	7.69	8.59
400	7.03	5.20	5.66	6.14
500	5.62	4.14	4.43	4.69
600	4.67	3.42	3.60	3.75
800	3.45	2.50	2.56	2.60
1000	2.70	1.93	1.94	1.94
3000	0.63	0.43	0.41	0.42
5000	0.25	0.17	0.16	0.17
10000	0.00	0.00	0.00	0.00

Note: Calculated for the $JHKL$ filters used in the SAAO (Carter, 1990) photometric system and a hypothetical N-band extending from 8 to $14\,\mu$m. The J-band is assumed to have a sharp cutoff at $1.37\,\mu$m. The color zero-points have been adjusted to be 0.0 for a 10000 K blackbody. The precise color values will differ from system to system (see Section 3.1.4).

1.2.7 Laboratory blackbodies

In a laboratory a "blackbody" calibration source consists of a small oven with walls at a carefully controlled temperature. Of course, there has to be an aperture somewhere in the wall to allow for the emission of some part of the radiation. The size of the aperture is kept small compared to the total surface area of the cavity, and various shapes have been devised which yield radiation at a pre-determined fraction of what a true blackbody would give at a particular temperature (Treuenfels, 1963). The output beam is uniform over only a small solid angle. See Wolf and Zissis (1978) for many practical details.

1.3 Atomic spectra

In this section a qualitative description of atomic spectra is given in order to provide a "feel" for the considerations involved and the meaning of the terminology. More rigorous accounts will be found in quantum mechanics and spectroscopy textbooks.

A spectral line arises by a transition from one energy state of an atom, E_1, to another, E_2, and its frequency is given by

$$h\nu = E_1 - E_2.$$

The energy level of the atom is the sum of many different contributions, for example:

- the translational kinetic energy of the atom as a whole (which affects the wavelengths of the spectral lines through Doppler shifts)
- the energy of the electrons in their orbits around the nucleus
- exchange forces between electrons which favour parallel spin vectors
- the mutual electrostatic interactions of the electrons, related to their angular momenta
- the interaction between the magnetic moments of the electrons with the magnetic field caused by their movement (as charges) around the nucleus (fine structure; *L-S* coupling)
- the interaction of the magnetic moment of the nucleus with the magnetic field just mentioned (hyperfine structure)
- the interaction of the magnetic field of the atom with an external magnetic field, if present (Zeeman effect)
- relativistic corrections

1.3.1 Hydrogen atom

In the simplest approach to the theory of the hydrogen atom, the energy level is taken to be the sum of the kinetic energy of the electron and its potential energy in the field of the positive nucleus. Only certain orbits are permitted according to quantum theory. They are characterized according to the integer n, the *principal quantum number*, and have energies proportional to $1/n^2$. When the atom changes state *radiatively*, a photon is absorbed or emitted having a *wave number* ω (number of wavelengths per cm) given by

$$\omega = R \left(\frac{1}{n_1^2} - \frac{1}{n_2^2} \right)$$

where n_1 and n_2 are integers, $n_2 > n_1$ and R is the Rydberg constant, 109678 cm^{-1}. The case $n_1 = 1$ corresponds to the Lyman (ultraviolet) series, $n_1 = 2$ to the Balmer (visible), $n_1 = 3$ to the Paschen, $n_1 = 4$ to the Brackett and $n_1 = 5$ to the Pfund series etc., named after their discoverers. As n_2 approaches infinity, the wave numbers converge to the *series limit* for each n_1 (Table 1.3).

An atom can also change its energy level through a collision with an electron, called *collisional excitation*, without the emission or absorption of a photon. This process becomes important in regions of space where the electron density $n_e > 10^4$ cm^{-3}.

Table 1.3. *Wavelengths of some hydrogen lines*

Series	Line	Wavelength(μm)
Lyman	Lyα	0.1216
($n=1$)	Lyβ	0.103
	Lyγ	0.0973
	Ly$_{\text{limit}}$	0.0912
Balmer	Hα	0.656
($n=2$)	Hβ	0.486
	Hγ	0.434
	H$_{\text{limit}}$	0.365
Paschen	Pα	1.876
($n=3$)	Pβ	1.282
	Pγ	1.094
	P$_{\text{limit}}$	0.820
Brackett	Brα	4.052
($n=4$)	Brβ	2.626
	Brγ	2.166
	Br$_{\text{limit}}$	1.459
Pfund	Pfα	7.460
($n=5$)	Pfβ	4.654
	Pfγ	3.741
	Pf$_{\text{limit}}$	2.279

This first-order solution also includes quantization of the angular momentum l (quantum number l, which can have integral values from 0 to $n-1$). The z-component of angular momentum is also quantized (quantum number m, which can have values from $-l$ to $+l$). Neither the angular momentum state nor its z-component affect the energy levels in the simplest form of the theory. The real atom, however, does show such a dependence on l. In addition, if an external magnetic field is imposed, the energy also depends on m.

To make the model of the H atom more realistic, the much smaller energy of the interaction between the magnetic moment of the electron (its spin angular momentum is s; quantum number s, which can have the value $-1/2$ or $+1/2$) and the magnetic field of the current generated by its orbital movement must be taken into account (*l-s* coupling). The different possible angular momenta then cause the energy levels to be split, giving rise to the observed *fine structure*. For hydrogen, the splitting of the energy levels is only of order 10^{-5} of the separation of levels with different principal quantum number n.

The different angular momentum states 0, 1, 2, 3, ... are denoted by the letters S, P, D, F, ..., corresponding to the names sharp, principal, diffuse,... that characterize the series of spectral lines they give rise to (mainly in more complicated atoms than hydrogen, which have more pronounced splitting).

Consideration of interactions with the even smaller nuclear magnetic moment leads to the *hyperfine structure*. The case of the hyperfine structure of the ground state of hydrogen is particularly important as transitions between the two levels generate the 1420 MHz line used in radio astronomy as a tracer of neutral hydrogen atoms.

1.3.2 Multi-electron atoms

In many multi-electron atoms, *L-S* coupling applies. The condition for this is that the spin-orbit energy must be smaller than the energies related to spin and angular momentum. A particular state is described by the quantum numbers for the individual electrons (small letters) together with those of the atom as a whole (large letters):

$$(n_1, l_1, s_1), (n_2, l_2, s_2), \ldots, L, S, J, M_J$$

where L, S, J are the quantum numbers for the total orbital angular momentum **L**, spin **S** and overall angular momentum **J**, respectively. M_J is the quantum number of its total z-component. **L** and **S** add vectorially, so that

$$\mathbf{J} = \mathbf{L} + \mathbf{S}.$$

Thus, for example, if $S = 1$ and $L = 2$, J can have the values 1, 2 or 3, according to the mutual orientation of **L** and **S**.

1.3.2.1 Shell structure

The relevant quantum numbers are said to define an *energy state, quantum state* or *term*. Going through the periodic table, as the atomic number increases we find that atoms are constructed as a series of *shells* and *subshells* of electrons. This is a consequence of the *Pauli exclusion principle* which states that each electron in the system must have a unique set of quantum numbers (n, l, m, s).

The first shell, corresponding to $n = 1$, is called the K shell. Since for $n = 1$, l can only be 0, there can only be at most two electrons in this shell. The one-electron case is hydrogen, and the two-electron case is helium, whose electrons have $s = 1/2$ and $-1/2$. The next shell (L) has electrons with $n = 2$, so that l can have value 0 or 1. The $l = 0$ state has the lowest energy, so that lithium, with atomic number Z = 3, has a complete K shell and one extra electron in the L shell, with $l = 0$. Be

($Z = 4$) has two electrons in the K shell, with $l = 0$ but opposite spin
s. The $l = 1$ *subshell* can have m values of -1, 0 or $+1$, so that it may
contain up to six electrons, allowing for the two spin states for each set
of n, l, m.

Each l-value gives rise to a subshell. The sub-shells are denoted $1s$,
$2s$, $2p$, $3s$, $3p$, $3d$, $4s$, $4p$, $4d$, $4f$,... for $l = 0, 1, 2, 3, \ldots$, etc.

As the M shell fills up, it is found that the $3d$ subshell has a higher
energy level than the $4s$. The $3p$ subshell is completed with Ar ($Z = 18$),
and the next two atoms K (19, potassium) and Ca (20) have electrons
in the $4s$ subshell. The $3d$ levels fill up thereafter.

The description of an atom in terms of shell structure is written as,
for example,

$$1s^2 2s^2 2p \text{ B } (Z = 5)$$

where the superscripts indicate the number of electrons in the subshell.
The atom with full $1s$, $2s$ and $2p$ subshells is

$$1s^2 2s^2 2p^6 \text{ Ne } (Z = 10).$$

The outermost shell is called the *valence shell*, because the electrons
in that shell define the chemical properties of the atom.

1.3.2.2 Transitions

The states defined by the principal quantum number n are split into
finer levels by (in descending order of importance) the spin S, the orbital
angular momentum L and the total angular momentum J. The last effect
is caused by L-S coupling.

The state of the outer part of the atom before or after a transition
is given by its total electronic angular momentum quantum number L
using the S, P, D, F,... notation. The leading superscript denotes the
multiplicity of the level ($2S + 1$, where S is the total electron spin quantum number) and the following subscript is the value of J, the total
angular momentum quantum number.

For example, the transition of neutral calcium CaI at $2.263\,\mu$m is
denoted by

$$4f\,^3F_3^o - 4d\,^3D_2.$$

Here the outermost electron goes from the $4f$ to the $4d$ state, the atom
as a whole goes from a triplet F electronic angular momentum state to
a triplet D state, while the total angular momentum quantum number

goes from 3 to 2 and the parity of the waveform, denoted by the o superscript, goes from odd to even. The ground state of CaI is

$$1s^2 2s^2 2p^6 3s^2 3p^6 4s^2.$$

1.3.3 Selection rules, permitted and forbidden lines

The theory of the interaction of atoms with the radiation field shows that only certain combinations of states are *permitted* to radiate or absorb easily, as electric dipoles. Other combinations are not absolutely *forbidden* to radiate, but must do so as magnetic dipoles or as electric quadrupoles and such transitions are inherently less probable. This fact gives rise to *selection rules* which restrict how the initial and final sets of quantum numbers may differ. The most important of these are that

- only one electron may change state
- its l quantum number must change by one unit

  The change of the l quantum number is a consequence of the fact that the parity of the wave function must change in an electric dipole transition. Where necessary, the parity of a configuration is denoted by the superscript o (for odd).
- J changes so that $\Delta J = 0, \pm 1$

  When *l-s coupling* is involved we have also
- $\Delta L = 0, \pm 1$
- $\Delta S = 0$, i.e., there are no so-called *intercombinations*, or lines from one total spin state to another. Thus singlet–triplet transitions etc. are forbidden.

Under conditions where a low-density gas occupies a vast volume, such as often occurs in interstellar space, large column densities of atoms sometimes become trapped in a state from which there is no permitted (or dipole) downward transition available. In these cases, magnetic dipole or electric quadrupole transitions *are* seen to occur as their intrinsically low probabilities are offset by the sheer number of atoms in the trapped state. Lines originating in this way are indicated by [ ] and are referred to as *forbidden*; for example, numerous transitions of [FeII] are observed in the infrared.

The most famous example of forbidden transitions are those at 4959 and 5007 Å, due to [OIII] and originally believed to be due to a hypothetical element, nebulium, which had not been identified terrestrially.

Another important transition is that of the 2 ^2S level of HI which cannot decay by a permitted transition to the ground state 1 ^2S. It can, however, decay by *two-photon emission*, where the energies of the two photons must add up to that of the Lyα line.

1.4 Molecules

The energy levels of a molecule are determined mainly by the following, in order of decreasing energy:

- electronic energy – changes in n give ultraviolet or visible lines
- vibrational energy – changes in v only give infrared lines
- rotational energy – changes in J typically give millimetre wave lines

In general, a spectral line may involve changes in all three of these quantities.

1.4.1 Electronic states

The electronic ground state of a molecule is labelled by the letter X. Excited states which can be reached by dipole transitions from the ground state are labelled by A, B, C, etc. Small letters a, b, c, etc. are used to label states of different spin multiplicity from the ground state.

1.4.2 Angular momentum

The electronic angular momentum $\mathbf{L}$ and the spin angular momentum $\mathbf{S}$ add vectorially to form the total electronic angular momentum $\mathbf{J} = \mathbf{L} + \mathbf{S}$ as for atoms.

However, in a diatomic molecule, with an axis of symmetry joining the two nuclei, the electronic angular momentum $\mathbf{\Lambda}$ lies along that axis and is quantized with quantum number Λ. The states (term symbols) are $\Sigma, \Pi, \Delta, \Phi$, etc., corresponding to values of Λ of 0, 1, 2, 3, etc. As in atoms, when Λ is non-zero the energy states are affected by spin-orbit (L-S) or similar couplings, because the orbital motion of the electrons creates a magnetic field along the internuclear axis that interacts with the electronic spin (magnetic moment). The total electron spin is $\mathbf{S}$, which can be integral or half-integral. $\mathbf{S}$ defines the *multiplicity* of the level because there can be $2S+1$ values of M_S (also confusingly called Σ) the component along the symmetry axis. The total electronic angular momentum about the symmetry axis is thus $\Omega = |\Lambda + \Sigma|$. The diatomic

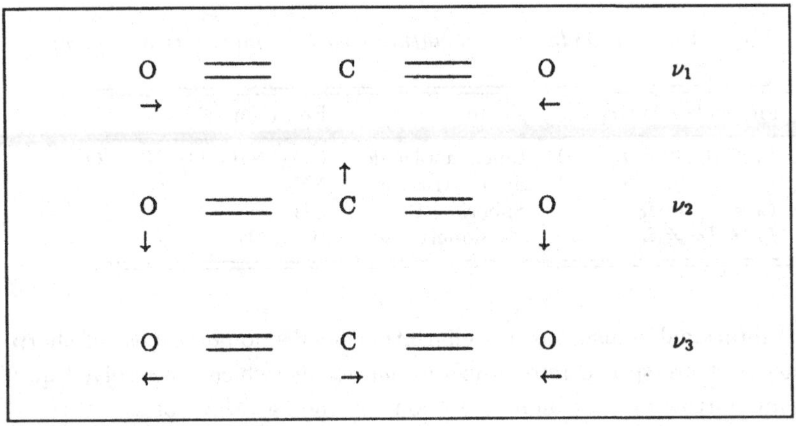

Fig. 1.1. Vibration modes of CO_2. ν_1 (wavelength 7.22 μm) and ν_3 (4.27 μm) are stretching modes and ν_2 (14.97 μm) is a bending mode.

electronic angular momentum state is thus specified $^{(2S+1)}\Lambda_\Omega$, although the Ω is often omitted. Two further symbols relate to the symmetry of the electronic wave function. The superscript $-$ or $+$ denotes, for molecules in the Σ state, whether the sign of the wave function changes when reflection takes place in a plane passing through both nuclei. The subscripts u (odd, *ungerade*) or g (even, *gerade*) denote whether the sign of the wave function changes with reflection in the origin, but only for molecules whose two atoms have the same charge.

1.4.3 Vibrational states

The nuclei of a diatomic molecule oscillate relative to each other in a potential which resembles that of a simple harmonic oscillator for the lower levels, but which ultimately allows dissociation as the energy increases. The state is indicated by the quantum number v. More complicated molecules can have extra modes of vibration, denoted v_1, v_2, v_3, etc.

For example, in CO_2, where the C atom is in the middle and the O atoms are at the opposite ends (see Fig. 1.1), there are three vibrational modes, two of which involve *stretching*. The other involves *bending*.

The energy of a vibrational level is given approximately as the sum of a series of simple harmonic oscillators:

$$E_V = (v_1 + 1/2)h\nu_1 + (v_2 + 1/2)h\nu_2 + (v_3 + 1/2)h\nu_3 + \cdots.$$

The state is specified as $(v_1, v_2, v_3, \ldots)$.

Table 1.4. *Molecular nomenclature based on moments of inertia*

Moments of inertia	Name	Representatives
$I_A = 0$, $I_B = I_C \neq 0$	Linear molecule	CO_2, N_2O, O_2, N_2, CO
$I_A \neq 0$, $I_B = I_C \neq 0$	Symmetric top	NH_3
$I_A = I_B = I_C$	Spherical top	CH_4
$I_A \neq I_B \neq I_C$	Asymmetric top	H_2O, O_3

Vibrational transitions in stellar atmospheres do not appear as sharp lines, but are spread into *ro-vibrational bands* (which are resolved into many narrow lines at high resolution) by the fact that rotational transitions occur at the same time. Solid molecular material – the ices in the denser parts of the interstellar medium – also show broad vibrational features, but the broadening has other causes since the molecules are no longer free to rotate.

1.4.4 Rotational states

The rotational energy is the least of the three major contributors to the energy level of a molecule. There are four types of rotating molecules, based on moments of inertia (see Table 1.4).

The rotational angular momentum is quantized with quantum number J.

1.4.4.1 Linear molecules

Dipole radiation is only possible if the molecule has a dipole moment – a *homonuclear* molecule such as H_2 has no dipole moment and can only radiate in less probable transitions. Similarly, CO_2 has no pure rotation spectrum. CO, on the other hand, has a pure rotation spectrum which is very important in the millimeter-wave region as a tracer of cool molecular gas. Its $J = 1-0$ and $J = 2-1$ lines, both in ^{12}CO and ^{13}CO, have been used extensively for mapping. The higher J-value lines occur in the far-infrared and are readily seen, for example, in the mass-losing carbon star IRC + 10216 (Cernicharo et al., 1996). The energy levels are proportional to $J(J + 1)$, with a small correction for stretching of the molecule. The pure rotational spectrum is thus a collection of almost equally spaced lines.

The nuclear symmetry can be classed as *a* or *s* (antisymmetric or symmetric) for linear molecules. The selection rule is

$$\Delta J = 0, \ \pm 1 \ (J = 0 \text{ to } J = 0 \text{ not allowed}).$$

1.4.4.2 Spherical tops

Because they lack symmetry axes, spherical tops have the same energy levels as linear molecules. They differ, however, in the statistical weights of the levels.

1.4.4.3 Symmetric tops

The component of angular momentum in the direction of the symmetry axis is now also quantized (quantum number K). K is integral, positive and $\leq J$. A pure rotational transition is specified by its initial and final K and J numbers. Symmetric tops are *prolate* if A > B and *oblate* if A < B. The energy levels are proportional to

$$BJ(J+1) + (A-B)K^2$$

where A and B are called the *rotational constants* and are inversely proportional to I_A and I_B, the moments of inertia mentioned in Table 1.4.
   The selection rule is

$$\Delta K = 0; \ \Delta J = 0, \pm 1 \ (J = 0 \text{ to } J = 0 \text{ not allowed}).$$

1.4.4.4 Asymmetric tops

These are intermediate between oblate and prolate symmetric tops. K ceases to be a useful quantum number. A state with rotational quantum number J has $2J+1$ energy levels associated with it.
   The selection rule is

$$\Delta J = 0, \ \pm 1 \ (J = 0 \text{ to } J = 0 \text{ not allowed}).$$

Here, because the rotational energy is no longer a monotonic function of J, all three values of ΔJ can give positive energy jumps (absorption spectrum), and these are then called:

$$\Delta J = -1 \quad P\text{-branch}$$

$$\Delta J = 0 \quad Q\text{-branch}$$

$$\Delta J = +1 \quad R\text{-branch}$$

1.4.4.5 Λ-doubling, fine and hyperfine structure of rotational states

There are smaller contributions to the energy level of a molecule which cause spectral lines to split into closely spaced components. For example, we can consider the interesting case of OH.

Firstly, the electrons in a real linear molecule cause the moment of inertia along the internuclear axis to have a non-zero value, effectively making it into a symmetric top molecule whose I_A is very small compared to I_B $(= I_C)$. Changes in the electron configuration affect I_A which, it turn, affects the rotational energy levels. If $\Lambda = 0$, no splitting occurs, but if $\Lambda \neq 0$, there will be two energy states of opposite symmetry according to the direction of rotation of the electrons around the axis. This is called Λ-*doubling*.

Secondly, fine structure is caused by the interaction of the electron spin magnetic moment **S** with fields caused by the electronic angular momentum and the molecular rotation. The overall angular momentum **J** is quantized and J can be an integer or an integer $+1/2$.

Lastly, hyperfine structure is due to the interaction of the nuclear magnetic moment with the other fields. The energy levels are again split according to whether the nuclear spin adds to or subtracts from the overall total angular momentum, **F**.

See Herzberg (1971) for a more complete discussion of these terms.

The lowest (ground) state of OH has $J = 3/2$, with term $^2\Pi_{3/2}$. It is split first by Λ-doubling and again by nuclear spin. The fine structure (electronic spin) interaction is such that the $J = 1/2$ $^2\Pi_{1/2}$ state has higher energy than the ground state.

Transitions between the four levels of the ground state of OH give rise to the OH lines seen around 1612 MHz in the radio. The levels are populated by infrared photons at 35 and 53 μm which excite the molecule to higher rotational levels which then decay to the ground state. Non-equilibrium population of the four levels leads to maser action in circumstances such as those that occur in the outer atmospheres of young stellar objects, compact HII regions and asymptotic giant branch variables.

1.4.5 Example 1: the water molecule

Absorption by water vapor, an asymmetric top molecule, plays an important role in infrared astronomy, both in the atmosphere of the Earth, through which most observations must be made, and in the atmospheres of cool stars, which are frequently dominated by water-vapor features.

1.4.5.1 Ortho- and para-water

Because the spins of the two hydrogen atoms of water can be parallel (ortho) or opposite (para), there are two species of water, occurring normally in the ratio 3:1.

1.4.5.2 Rotation-vibration spectrum of water

The vibrational modes of H_2O are three in number. The lowest frequency (cm^{-1}) mode is the ν_2 which gives a band at $1595 \, cm^{-1}$ ($6.25 \, \mu m$). The others are the ν_1 and ν_3 which give bands at 2.74 and 2.66 μm. The absorption bands which dominate the atmospheric transmission in the 1–$2 \, \mu m$ region (see Fig. 2.1) are hybrids of two or more transitions.

The complexity of the water vapor bands is such that experimental measurements of the band intensities are usually resorted to.

Measurements of the absorption coefficient of water vapour $k(\omega, T)$ have been made by a number of investigators, for example, Ludwig (1971). The absorption coefficients given (Fig. 2.2) are those for the idealized case of an optically thin absorber. The resolution of Ludwig's work is $25 \, cm^{-1}$. Since the actual absorption within any given finite band will be the result of many individual lines, broadened by natural and pressure effects, the relationship between absorption and path length will not be a simple one, i.e., Lambert's law (see Section 2.2) is not obeyed. It is necessary to model the effect of an absorber of finite optical thickness to take care of this problem.

1.4.5.3 Pure rotation spectra of water

Many pure rotational lines of both ortho- and para-water vapor have been detected in W Hya, an O-rich Mira, by Neufeld et al. (1996) and Barlow et al. (1996).

1.4.5.4 Band models

To cope with studying absorption over finite bandwidths, much greater than the width of individual spectral lines, various *band models* have been introduced.

For example, the statistical model for collision-broadened lines assumes that the individual line intensities may be described by an exponential probability function

$$p(S) = \frac{1}{\sigma} e^{-S/\sigma}$$

where $p(S)dS$ is the probability that a line has intensity between S and

$S + \mathrm{d}S$ and σ is the mean line intensity, and the positions of the lines are assumed to be randomly located. This model has been found to give a good representation of measured spectral absorptivities and emissivities at high temperatures.

To obtain the average absorption coefficient $k(\omega, T)$, where ω is the wave number in cm^{-1} in each wavelength interval, Ludwig (1971) also had to solve for the fine-structure parameters $a(\omega, T)$. Measurements were taken for a range of pressures, temperatures and path lengths.

Ludwig's results for two temperatures are shown in Fig. 2.2. It will be noticed that the absorption bands are much narrower at 300 K and that the transmission bands are much more transparent than at higher temperatures.

Since red stars have effective temperatures around 3000 K, many of their water-vapor bands can be observed from the ground in spite of the water content of the Earth's atmosphere. In fact, it is possible to determine their temperatures by observation of their water-vapor band characteristics.

1.4.5.5 Rigorous treatment of water-vapor lines

Recent stellar models have been able to take advantage of increasing computer power to treat the individual spectral lines of various molecules, as listed in massive data banks. As many as 10^6 lines of HCN and H_2O are known (Jørgensen, 1997).

1.4.6 Example 2: molecular hydrogen

The electronic ground state of the H_2 molecule is $X\,^1\Sigma_g^+$. The X means that it is the ground state, the 1 that the multiplicity is 1 (i.e., S = 0), the Σ that the component of electronic angular momentum along the symmetry axis is 0, etc.

Because molecular hydrogen has no dipole moment, it is forbidden to make dipole transitions. However, under certain conditions it becomes a very important absorber or emitter. Two important instances are in a photodissociation region (PDR) in the interface between molecular cloud and HII regions and in the atmospheres of very cool high-gravity stars such as M dwarfs and the "brown dwarfs".

Quadrupole radiation, with a transition probability of about 10^{-9} times that of a typical permitted dipole, is possible. In fact, H_2 is frequently observed wherever molecular clouds are sufficiently heated by collision or non-thermal processes.

There are 15 bound vibrational energy levels, identified by the quantum number ν. Rotational energy levels associated with each vibrational level are denoted by the quantum number J (Sternberg, 1990).

Because of the quadrupole nature of the transitions, the rotational quantum number must change by 2, 0 or –2 (except 0–0) with no restrictions on the vibrational quantum number. The notation for a particular line gives the upper vibrational quantum number, followed by the lower one and one of S(J), Q(J) or O(J), where J is the rotational quantum number of the lower level and S, Q and O denote whether J is smaller than, equal to or greater than the rotational quantum number of the upper level. For example, the line at $2.12\,\mu$m is due to the 1–0 S(1) transition, where the vibrational quantum number goes from 1 to 0 and the rotational quantum number (J) goes from 3 to 1.

1.5 Synchrotron radiation

Synchrotron radiation is emitted by a charged particle spiralling in a magnetic field. Continuum radiation from astrophysical objects is often observed to have a power-law spectral distribution. Though this is usually attributed to synchrotron radiation from relativistic electrons, it is also possible to generate a power law in other ways.

If the spiralling particles have a power-law energy distribution with exponent γ, so that

$$n(E)\mathrm{d}E \propto E^{-\gamma}\mathrm{d}E$$

it can be shown that the emitted electromagnetic spectrum will also have a power-law distribution, but with exponent

$$\alpha = -(\gamma - 1)/2,$$

where $F_\nu \propto \nu^\alpha$. *Synchrotron self-absorption* occurs in intense synchrotron sources at frequencies below

$$\nu_{\mathrm{cutoff}} \sim 20n_e/B$$

where n_e is the electron density and B is the magnetic field in cgs units.

Synchrotron radiation is usually highly beamed because of relativistic effects; the opening angle ϕ is given by

$$\phi \sim (1 - v^2/c^2)^{1/2}$$

where v is the velocity of the radiating electrons. However, a tangled

Table 1.5. *Colors of power-law spectral distributions*

Exponent	$J - H$	$H - K$	$K - L$	$L - N$
−2	1.11	1.13	1.80	4.63
−1	0.78	0.82	1.31	3.39
0	0.46	0.50	0.83	2.19
1	0.13	0.18	0.34	1.00
2	−0.20	−0.14	−0.14	−0.16

Note: Calculated for the filter characteristics of the SAAO (Carter, 1990) system and an N-band which transmits from 8 to $14\,\mu$m. The zero-points of color have been set so that a $10000\,$K blackbody has color zero in each index. The precise values of the colors will differ from system to system (see Section 3.1.4).

magnetic field can reduce the degree of beaming and many sources have sufficiently tangled fields that their emission is essentially isotropic.

The colors of various power-law distributions are given in Table 1.5.

1.6 Further reading

Hartquist, T. W. (ed.), 1990. *Molecular Astrophysics; A Volume Honoring Alexander Dalgarno*, Cambridge University Press, Cambridge.

Herzberg, G., 1945. *Molecular Spectra and Molecular Structure. II Infrared and Raman Spectra of Polyatomic Molecules*, van Nostrand Co., New York. A "classic" text.

Herzberg, G., 1971. *The Spectra and Structures of Simple Free Radicals, An Introduction to Molecular Spectroscopy*. Cornell University Press, Ithaca, New York; reprinted by Dover Publications Inc., New York, 1988.

Jørgensen, U. G. (ed.), 1994. *Molecules in the Stellar Environment*, Proc. IAU Colloquium 146, 1994, Springer, Berlin, p. 29.

Osterbrock, D. E., 1989. *Astrophysics of Gaseous Nebulae and Active Galactic Nuclei*, University Science Books, Mill Valley, California.

2

The Infrared Sky

2.1 Introduction

As we examine the sky at longer and longer wavelengths, the objects that appear most prominent generally have lower and lower temperatures. This is because a blackbody distribution is a useful model for many celestial objects and the Wien displacement law (see section 1.2.2) states that

$$T\lambda_{\text{peak}} = 2898.$$

For example, the effective wavelength of the V-band ($0.55\,\mu$m) corresponds to the maximum of the eye's sensitivity in bright sunlight and is close to the wavelength of maximum spectral radiance of the Sun, whose effective temperature is $\sim$6000 K. In comparison, the coolest normal stars ($\sim$3000 K) peak at around $1\,\mu$m. Dust is often found at a temperature of a few hundred degrees, peaking at around $10\,\mu$m. The relic of the Big Bang, the 3 K cosmic background, peaks at about $1000\,\mu$m, just beyond the end of the infrared and the beginning of the radio.

The infrared sky in the very near, or photographic, infrared shows few surprises and is usually treated as an extension of visible-region astronomy because of the similarity of the techniques that are used. Parts of the sky, mainly near the galactic plane, have been surveyed in the "photographic I" band, at around $0.8\,\mu$m, by the Palomar and UK Schmidt telescopes using IV-N plates (see Hoessel et al., 1979). The I-band offers greater penetrating power through interstellar dust than B and V and is also great of interest in searches for cool stars (including variables). Although limited fields have been surveyed in the "photographic Z" band at around $1\,\mu$m, the small increment in wavelength fails to offset the reduced sensitivity of the available plates.

The infrared proper, technologically speaking, starts at the point ($\sim 1.1\,\mu$m) where many of the visible-region technologies such as photography, intrinsic silicon detectors (CCDs) and photocathodes cease to be useful. This is somewhat beyond the point at which the average human eye loses all sensitivity (about $0.75\,\mu$m).

By $2.2\,\mu$m (K-band) the sky itself takes on a significantly different appearance from the visible one. The cooler end of the stellar population becomes predominant and some objects turn out to have infrared fluxes much higher than predicted from their visible spectra. These *infrared excesses* can be due to cool dust shells surrounding them or to circumstellar free-free emission, for example.

Beyond about $2.3\,\mu$m blackbody radiation from the telescope and the atmosphere itself begins to dominate other sources of background. Measurements of faint astronomical objects have to be made by alternatively observing the field containing the source and a nearby "empty" one. This process is known as *chopping*. The signals are subtracted to eliminate the strong background.

At $10\,\mu$m the blackbody radiation from ~ 300 K terrestrial sources reaches its maximum. At these long wavelengths, broad-band array detectors must often be read out several tens of times per second because they rapidly become saturated with background photons. The signals are digitally processed at high speed to reveal the relatively faint astronomical sources.

As the mirrors and support structures of the telescope radiate strongly, attention has to be given to reducing the background they cause. The reflective optical surfaces can be coated with gold instead of the usual aluminium to reduce their infrared emissivity. There is no possibility of cooling the optics and support structure of a ground-based telescope since this would simply cause them to become coated with ice from the water vapor always present in the air.

By placing a cooled telescope in space, all the problems associated with the transmission and emission of the Earth's atmosphere may be overcome and the telescope itself may also be cooled. Zodiacal emission and "infrared cirrus" then set the background level (see section 2.4).

Not all infrared wavelengths are equally suitable for ground-based observations because the transparency of the Earth's atmosphere is strongly variable with wavelength (Fig. 2.1). The first deep band of absorption occurs around $0.93\,\mu$m and others cut out smaller or larger wavelength regions until, at about $26\,\mu$m, the sky effectively becomes opaque. Limited transmission bands may sometimes occur to about $40\,\mu$m when the

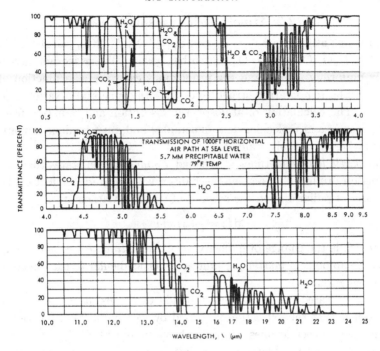

Fig. 2.1. Transmittance of 1000 feet horizontal air path at sea level containing 5.7 mm of precipitable water at 79 F (26 C). The omitted interval from 9.5–10 μm has transmittance = 100%. (Image is courtesy of Raytheon Infrared Center of Excellence, formerly SBRC.)

atmospheric water-vapor content is exceptionally low. Beyond wavelengths of several hundred microns, bands of partial transmission again become available and by millimeter wavelengths various quite clear bands are found.

The "end" of the infrared region is again technologically defined: it is the wavelength ($\sim$350 μm or 0.35 mm) where radio techniques such as superheterodyne receivers tend to be used in preference to the "optical-style" infrared approach and the incoming radiation tends to be thought of as waves rather than individual photons. The region from 0.35 mm to 1 mm is referred to as the *sub-millimeter region*. It may be regarded as a sub-division of radio astronomy.

The convention has grown up of dividing the infrared as a whole into the near- (0.75 to 5 μm), mid- (5 to 25 μm) and far-infrared (25 to 350 μm).

2.2 Atmospheric transmission

The Earth's atmosphere is composed of N_2, O_2, A, CO_2, Ne, He, CH_4, Kr, H_2, N_2O, CO, H_2O, O_3 and other gases. These are not necessarily uniformly mixed, and the proportions can vary as a function of height. The temperature structure of the atmosphere also varies with height. The calculation of atmospheric absorption and radiation from basic atomic and molecular data is complex and must be done numerically rather than analytically.

Lambert's law (sometimes also called *Bouguet's law*) states that the extinction produced by a given mechanism is linear, independent of the intensity of radiation and the amount of matter. Mathematically,

$$dI_\nu = -e_\nu I_\nu da$$

where dI_ν is the change of intensity, e_ν is the *extinction coefficient* and da is the amount of matter per unit area perpendicular to the direction of propagation.

In the infrared, e_ν is a strong function of the frequency ν because of the fact that the extinction is caused by vast numbers of individual atomic and molecular lines. Each line has a finite width, which is partly intrinsic (due to the lifetime of the initial and final states) and partly due to circumstances, especially pressure broadening. So long as the frequency band being considered is a small fraction of the width of the line, Lambert's law holds. However, many measurements are made over bandwidths which are much wider than the width of a single line, and Lambert's law is then no longer applicable. At some frequencies within a band, an atmosphere may be almost totally opaque, while at others it may be perfectly transmitting.

For this reason, the transmission must be calculated laboriously as a function of frequency using closely spaced intervals, and also be carried out for each path length of interest. However, as we shall see, some generalization is possible.

Figure 2.2 shows the transmission of water in the infrared (Ludwig, 1971). It is clear by comparison with Fig. 2.1 that most of the atmospheric absorption is due to H_2O. The main additional absorber is CO_2.

2.2.1 Theoretical atmospheric transmission

A classical treatment of the Earth's atmosphere from the infrared astronomer's point of view is that of Traub and Stier (1976) who used a

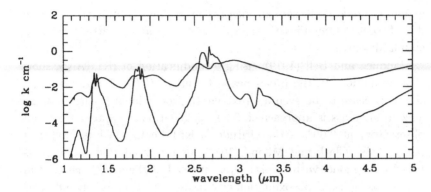

Fig. 2.2. The lower curve is the experimentally determined absorption coefficient of water vapor in the infrared at 300 K, appropriate to the terrestrial atmosphere. The upper curve is for 3000 K, appropriate to cool stars; based on data from Ludwig (1971).

computer programme known as IRTRANS. In this model, some 109,000 transitions of H_2O, O_3, O_2, CO_2, CO, N_2O and CH_4 molecules are taken into account.

They used the *Curtis Godson* approximation which assumes that a single line absorbing along a non-homogeneous path can be simulated by a line absorbing along a homogeneous path, with a single temperature and amount of absorbing matter. This approximation is shown to give realistic results, by comparison with experimental data. Wavelengths of 1 to 1000 μm were covered, and conditions corresponding to the altitude of the Mauna Kea observatories, aircraft and high-altitude balloons.

The thickness of atmosphere through which rays have to pass is measured in terms of *air mass*, where one air mass is the thickness of the atmosphere when looking straight up. It is given fairly accurately by sec Z, where Z is the zenith angle. In broad-band visible photometry, it is usually adequate to assume that the extinction due to the atmosphere in a particular band is given by

$$E_\lambda = C_\lambda \sec Z$$

where C_λ is a constant appropriate to the band in consideration. However, although the same formula is normally used in the infrared for comparisons between standard stars and measured objects when the difference in values of sec Z is small, it is not sufficiently accurate for work of the highest precision. It is particularly inappropriate to simply

extend the relation to zero air mass, when trying to obtain the absolute
flux of an object outside the atmosphere by comparison with a terrestrial
standard source.

Manduca and Bell (1979), using a modification of IRTRANS, show
that extinctions in the broad near-IR bands (*JHKL*, 1–5 μm; see section 3.2) cease to be even approximately proportional to the air mass
as zero air mass is approached. Their calculation is appropriate to an
observatory at 2.06 km, the altitude of Kitt Peak, which is typical of
many. Volk, Clark and Milone (1989) later showed that calculations
based on a model with several layers instead of the single one used in
the Curtis Godson approximation give results which are barely different.

The Manduca and Bell work covered the computation of extinctions
for summer and winter conditions, air masses from 0.5 to 3.0, filters
of different widths and stellar energy distributions for a Vega (hot star)
model and a cool giant model. They showed that Lambert's law is poorly
obeyed for air masses less than one and that even between 1 and 2 air
masses (the range in which most astronomical photometry is done) small
but significant errors in photometry can arise. For example, if early-type
standard stars are used with filters similar to the "Kitt Peak" set shown
in Fig. 3.1, errors of 0.02 mag may occur when red stars are measured
during the high water-vapor conditions prevailing in summer. Because
the problem is caused by increasing numbers of absorption lines near the
edges of the broad-band filter bandpasses it is even worse for the older,
wider, Johnson filters.

Absolute calibration of infrared broad-band photometry by comparison with ground-based standard sources is clearly made very difficult by
the lack of validity of Lambert's law. As a result, most photometry is
done by comparison of unknown objects with known ones which are also
outside the Earth's atmosphere (*standard stars*). Ideally, standard star
observations are made as close as possible in air mass to those of the
unknown object and should bracket it in time. To obtain the highest
accuracy, the spectrum of the standard star should be similar to that of
the unknown.

2.2.2 Water vapor

The water-vapor content of the atmosphere varies strongly with temperature and hence season. Angione (1989) shows that the column density
(usually expressed as "precipitable water in mm") measured at Mount

Laguna Observatory (altitude = 1860 m; latitude 33 N) varies from below 1 mm to over 15 mm, with around 4 mm being typical. The annual
average value also varies considerably.

The atmospheric pressure below about 120 km altitude is approximately given by $p_h = p_0 e^{-h/H}$ where h is the altitude and H is the scale
height (height at which pressure drops by factor e) which is around 8 km,
depending on temperature. The water-vapor content, however, falls off
much more rapidly with height, so that there is considerable advantage
in having observatories at high altitude. Traub and Stier (1976) suggest that the scale height of water vapor above Mauna Kea, Hawaii, is
1.85 km.

2.2.3 Scattering

Scattering from air molecules and aerosols play a role in infrared absorption by the atmosphere. The air molecules give rise to Rayleigh
scattering which is proportional to $(1/\lambda)^4$. Angione (1989) quotes a figure of only 0.003 mag per air mass at J (1.25 μm), arising from this
source, at Mount Laguna Observatory.

Mie scattering from aerosols such as sea salts, hydrocarbons and volcanic dust is also often a significant component of atmospheric extinction
according to Angione (1989). Its wavelength distribution depends on the
size distribution of the particles causing the extinction. The quantity
of aerosols in the atmosphere can vary greatly, increasing dramatically
after a massive volcanic eruption such as El Chichon, which occurred
in 1982. Measured (at 1.01 μm) and extrapolated values for the aerosol
components of the JHK extinctions at Mount Laguna Observatory as
given by Angione (1989) are shown in Table 2.1.

Wind-blown dust from the Sahara, such as blows across the Canary
Islands, consists of relatively large (micron-sized) predominantly silicate
particles. Calculations by Whittet, Bode and Murdin (1987) show that,
while the extinction should be predominantly neutral at visible wavelengths, it will show a strong decrease with wavelength in the infrared.

Aerosols are found very high up in the atmosphere, so that a site such
as Mauna Kea is not necessarily significantly better than a lower one,
so far as this source of extinction is concerned.

2.2.4 Variability of extinction

Glass and Carter (1989) have measured extinctions at Sutherland, which
lies at 1800 m altitude and latitude 32 S, and may be regarded as typifying

Table 2.1. *Measured and extrapolated extinctions caused by the aerosol component, in magnitudes per unit air mass*

Value	1.01 μm	J	H	K
Max	0.057	0.046	0.035	0.026
Min	0.005	0.004	0.003	0.002
Mean	0.020	0.016	0.012	0.009

After the eruption of El Chichon

Max	0.152	0.123	0.093	0.068
Min	0.036	0.029	0.022	0.016

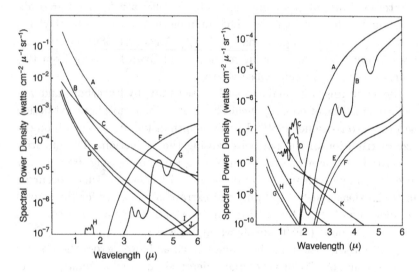

Fig. 2.3. *Left:* Day sky background. A–E, scattered sunlight for various altitudes and conditions; F is a blackbody at 283 K; G is emission of water vapor and CO_2; H = bright aurora; I,J = haze radiance + scatter of Earth flux under different conditions. *Right:* Night sky background. J, city lights; A, blackbody at 283 K; B, emission of water vapor and CO_2; C, aurora; D, airglow; E, F, haze and scatter of Earth flux under different conditions; G–K, scattered moonlight for various conditions; from Stewart and Hopfield (1965).

medium-latitude sites at moderate elevations (such as Kitt Peak, Palomar, La Palma and several of the Chilean observatories). The short-term averaged zero-points of the infrared colours J–H and K–L show clear variations with time of year. The J- and L-bands are more

affected by water vapor than the other two, the extinction being lower
in the affected bands in winter, when the temperature is low and the
atmosphere cannot hold so much water. The $H-K$ zero-point is surprisingly constant. There is no obvious seasonality in the behavior of the
$UBVRI$ extinction (J.W. Menzies, private communication). Zero-point
variations (in magnitude) of a few hundredths during a single night are
not uncommon in the infrared. The situation is however acceptable for
practical photometry because observations usually are made between
1.0 and 1.4 air masses, with standard star measurements every hour or
two.

Because the measurement of extinction requires observations at both
low and high air masses and is generally time-consuming, they are not
measured routinely. However, E_J, E_H and E_L have been plotted by
Glass and Carter against E_K for a number of nights. E_K ranges from
0.055 to about 0.18 mag air mass^{-1} and is typically about 0.10. E_H is
typically 0.06; E_J is 0.10 and E_L is 0.2 mag air mass^{-1}.

2.3 Terrestrial background radiation

The night time sky background is dominated at wavelengths less than
about $2\,\mu$m by Rayleigh scattered moonlight and starlight, to which are
added mainly lines of OH from the airglow (Fig. 2.3).

Beyond about $2.3\,\mu$m the essentially blackbody thermal radiation of
the atmosphere, modified by bands of partial transparency, dominates
the emission. During daytime, scattered sunlight dominates the thermal
radiation to somewhat longer wavelengths ($\sim 4\,\mu$m).

The structure of the telescope also causes a long-wavelength blackbody background. By using clean, gold-coated reflecting elements the
emissivity of the surfaces can be kept to very low levels. Cold baffles
(for example a Lyot stop; see section 6.3.2.5) are used to mask out warm
zones in the exit pupil of the telescope.

2.3.1 Airglow

The airglow represents the dominant source of background for broad-
band observations at J-, H- and much of the K-($2.2\,\mu$m) bands. It is
mainly radiated by excited levels of the hydroxyl radical OH$^-$, known
as the Meinel bands. Excitation is through the reaction

$$H + O_3 \rightarrow OH^* + O_2.$$

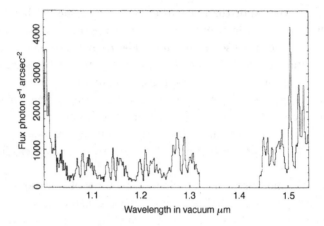

Fig. 2.4. Airglow emission in the *J*-band; from Ramsey, Mountain and Geballe (1992). The region around 1.4 μm was not measured.

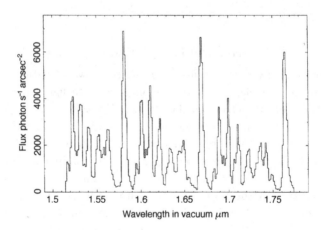

Fig. 2.5. Airglow emission in the *H*-band; from Ramsey, Mountain and Geballe (1992).

The altitude of the airglow emission is about 85–100 km, so that it affects all terrestrial sites. It is expected to be less at sites nearer the magnetic equator than the magnetic poles.

Tables of relative line intensities are given by Ramsay, Mountain and Geballe (1992) for observations made on a particular date from Mauna Kea. The lines are numerous and highly variable. They are most prominent in the *H*-band, though they also contribute to the *J*- and *K*-bands (Figs. 2.4–2.6). The sky brightness in the airglow lines varies on large angular scales and with time, due to the passage of gravity waves in

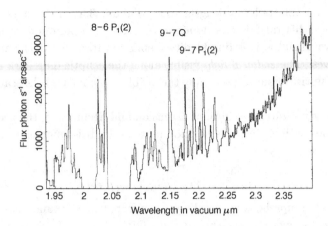

Fig. 2.6. Airglow emission in the K-band; from Ramsey, Mountain and Geballe (1992).

the ionosphere. The variations are characterized by spatial periods of tens of km and temporal periods of 5–15 minutes, with amplitudes of typically 10%. A decrease of intensity of order 50% during the night is not unusual.

The averaged integrated line flux in the H-band, amounts to about 3×10^4 photons s^{-1} m^{-2} arcsec^{-2} μm^{-1} (Maihara et al., 1993).

Several authors have attempted to find ways of suppressing the OH emission in order to obtain fainter limiting magnitudes, but no simple filter scheme has proved possible (e.g., Herbst, 1994). Most attempts have involved dispersing the incoming radiation and using a mask to eliminate the undesired lines. The spectrum may then be re-combined. This method can be used for building up a picture from a series of one-dimensional long-slit images. For example, Maihara et al. (1993) have proposed the construction of an imager which contains a spectroscopic filter to obtain a gain in the limit of 1.5–2 mag.

Oliva and Origlia (1992) point out that the lines can be used for wavelength calibration of spectra.

2.4 Extraterrestrial background sources

2.4.1 Zodiacal light background

The main extraterrestrial backgrounds with which infrared observers have to contend arise from the dust associated with the solar system, lying close to the plane of the planetary orbits (the *ecliptic*).

At nearer infrared wavelengths, out to about 3.5 μm, scattered sunlight from this dust is the dominant source of background, while at longer wavelengths it is due to direct emission from the dust particles themselves. The *zodiacal light* visible along the ecliptic on a dark night, around sunrise or sunset, is simply the visible component of the scattered light.

In any given direction in right ascension and declination, the zodiacal light varies with the Earth's position in its orbit. It is complex to model and subtract away.

2.4.1.1 The COBE satellite

The most comprehensive database on infrared backgrounds has been furnished by experiments aboard the COBE (Cosmic Background Explorer) satellite, in particular the DIRBE (Diffuse Infrared Background Experiment) and FIRAS (Far-Infrared Absolute Spectrophotometer). See Dwek (1995) for the proceedings of the COBE workshop on *Unveiling the Cosmic Infrared Background*.

DIRBE was designed to yield maps of the sky in bands centered at 1.25, 2.2, 3.5, 4.9, 12, 25, 60, 100, 140 and 240 μm, with a spatial resolution of $0.7° \times 0.7°$.

FIRAS is a Michelson interferometer providing cover from 100 μm to 10 mm wavelength, with a field of view having 7° FWHM.

Figure 2.7 shows the intensity of the night sky in terms of νI_ν, mainly based on COBE data.

2.4.2 Interstellar medium

Beyond about 100 μm, and until the cosmic microwave background becomes dominant, emission by the dust within the interstellar medium contributes more to the diffuse infrared background than the zodiacal light. This component is also known as *infrared cirrus*.

2.4.3 Cosmic microwave background

The cosmic microwave background (CMB) is the relic of the Big Bang. It is characterized by a blackbody spectrum of temperature 2.73 K and deviates from the blackbody distribution by less than 0.03% of its peak intensity between 500 μm and 5 mm wavelength.

The CMB becomes the dominant background at about 300 μm.

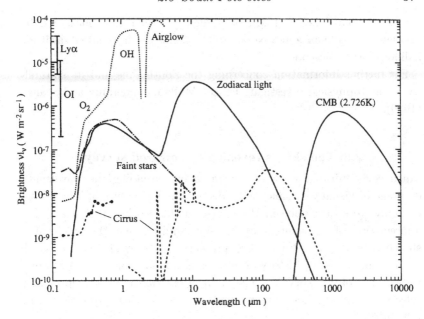

Fig. 2.7. Specific intensity ($\times \nu$) of diffuse emission from the night sky, observed away from the galactic and ecliptic planes, from high in the Earth's atmosphere; from Leinert et al. (1998). Mainly derived from COBE data.

2.5 South Pole sites

The altitude of the scientific base at the South Pole is 2800 m and the low surface temperature, typically -60 C in winter, makes it a very promising site for infrared studies. The precipitable water vapor is in the range 0.1 to 0.3 mm. A useful compendium of information about the Antarctic plateau is to be found in a report *Astronomy from the Antarctic Plateau* (Burton et al., 1994).

Ashley et al. (1996) show low-resolution spectrophotometry of the sky at the zenith from 2.0 to 2.55 μm and 2.9 to 4.2 μm at both the South Pole and Siding Spring, Australia. Below 2.2 μm the sky is dominated at both sites by airglow, but in the 2.3–2.5 μm window, where the airglow is minimal, the polar site has about two orders of magnitude less emission. In the L-band the sky brightness is 20–40 times less at the Pole. In fact, the real gain in observing at the Pole, where only small telescopes are available, is in the 3–5 μm region rather than at K.

Nguyen et al. (1996) make similar comparisons, pointing out that at a typical temperate latitude site the sky background at K is about 4000 μJy arcsec^{-2} ($K = 13.0$). At the pole, at 2.36 μm, it is about

$160\,\mu$Jy arcsec^{-2} ($K = 16.5$), only slightly higher than measured by Matsumoto, Matsuura and Noda (1994) at $2.38\,\mu$m from a high-altitude balloon-borne telescope.

For further information concerning the South Pole and its suitability for astrophysical investigations, see Mullan, Pomerantz and Stanev (1989).

2.6 The sky as revealed by infrared surveys

Much early infrared work concentrated on extending the wavelength coverage of visually conspicuous objects. The techniques of infrared astronomy were developed in the course of this work. However, a true appreciation of the content of the sky required that the whole of it should be surveyed. Price (1988) presents a "survey of surveys" which both summarizes the history of infrared astronomy and gives a comprehensive list of sky surveys. The surveys mentioned below represent a selection of those made or being made, chosen for their probable lasting influence.

2.6.1 Ground-based near-infrared surveys

2.6.1.1 The IRC survey

The IRC or, to give it its proper name, *The Two-Micron Sky Survey*, by G. Neugebauer and R.B. Leighton (1969), was the first systematic attempt to survey the sky at infrared wavelengths. Areas of sky below declination $-33°$ were not covered. The catalogue contains 5562 stars brighter than $K = 3.0$, with positions accurate to about 1 arcmin. The survey was conducted at I ($0.84\,\mu$m) as well as K ($2.2\,\mu$m) to give some idea of the color of each source.

Even with its relatively low sensitivity the IRC found some surprising results. In particular, it found a number of strong infrared sources that were faint in the visible. These included the young Becklin–Neugebauer object in the Orion nebula as well as evolved stars such as the luminous carbon star IRC $+10216$ which is obscured in the visible by a thick circumstellar dust shell.

2.6.1.2 DENIS

The DENIS project is surveying the southern sky from $-88°$ to $+2°$ with a 1 m telescope at ESO, Chile. It will observe simultaneously, using dichroic beam-splitters, the Gunn-i (0.82 μm), the J (1.25 μm) and the K_S (2.15 μm) or short K-bands. For a description see Epchtein et al. (1994).

2.6.1.3 2MASS

The Two-Micron All Sky Survey (2MASS) intends to survey the entire sky in the J-(1.25 μm), H-(1.65 μm) and K_S-(2.15 μm) or short K-bands between 1997 and 2000. It will use 1.3 m dedicated telescopes at Mount Hopkins, Arizona, USA, and Cerro Tololo, Chile. A description is given by Kleinmann et al. (1994). Like the DENIS survey, 2MASS makes used of NICMOS array detectors. See also http://www.ipac.caltech.edu.

2.6.2 Surveys from space

Large-scale surveys in the thermal infrared (beyond about 2.2 μm) cannot be conducted efficiently from ground-based observatories because of the unavoidable high backgrounds. With the arrival of the space-age, efforts were soon made to use rocket-launched, cooled, telescopes of small aperture to overcome this problem.

2.6.2.1 AFCRL and AFGL

The US Air Force Cambridge Research Laboratory (now Philips Laboratory) used a series of rocket-borne telescopes of 16.5 cm aperture to make surveys in three bands, 3–5 μm, 8–14 μm and 16–24 μm (Walker and Price, 1975). The AFGL catalogue (Price and Walker, 1976; Price, 1977; Price and Murdock, 1983) which was produced as the result of these and some other US Air Force surveys, showed up new classes of very cool objects such as proto-stellar sources, HII regions and proto-planetary nebulae.

2.6.2.2 The IRAS survey

The IRAS (infrared astronomical satellite) survey was performed from a 60 cm telescope in space which was cooled by liquid helium (Neugebauer et al., 1984). It functioned from January to November 1983, after which its helium supply ran out. It covered more than 96% of the sky in broad bands centered on 12, 25, 60 and 100 μm. Its results were published as catalogues of point sources, small extended sources, low-resolution spectra, as well as grey-scale images and various maps of the sky (see also infrared databases, below).

More than 245,000 point sources are listed in the resultant catalogue. A special issue of *Astrophysical Journal Letters*, 1984, **278**, L1–L85, was devoted to the first results.

As well as its photometric instruments, IRAS also carried a spectrometer which operated from 7.7 to 22.6 μm and measured 5425 sources brighter than 10 Jy with a resolution between 20 and 60 (Olnon, Raimond and the IRAS Science Team, 1986). More than 800 extra spectra were published by Volk and Cohen (1989) and Volk et al. (1991).

The main details of this satellite, its survey and its limitations were presented in Beichman et al. (1988).

A conference devoted to work done with IRAS was held in Noordwijk, the Netherlands, in 1985 (Israel, 1986).

2.6.2.3 IRTS

The Infrared Telescope in Space (IRTS) was a project of ISAS (Japan), launched in March, 1995, for a 28 day mission in which it surveyed ∼7% of the sky with four focal-plane instruments. (See Okuda et al., 1997, for information concerning this mission and its results.)

2.7 Balloon and airplane observatories

Balloon and airplane observatories operate above most of the water vapor of the atmosphere, but cannot achieve the low operating temperatures possible with cryogenic telescopes in space. They have, however, the advantage that flights are relatively cheap and the instrumentation can easily be changed.

2.7.1 SOFIA

The Stratospheric Observatory for Infrared Astronomy (SOFIA) is a US/German project to carry a 2.5 m telescope aboard a converted Boeing 747-SP airliner. It will fly at an altitude of 12.5–14 km, where the remaining atmosphere contains less than 5 μm precipitable water vapor. The temperature of the optics will be ∼240 K. The large aperture size gives improved spatial resolution at long wavelengths when compared to satellite-borne telescopes. The first scientific operations are expected in 2001 (Becklin, 1997). See also http://sofia.arc.nasa.gov.

2.8 Satellite observatories

Some recent infrared satellite projects have been designed to be used like ground-based observatories. They are not intended solely for surveys, but can be pointed at specific targets.

2.8.1 The ISO satellite

The ISO satellite of the European Space Agency was launched during November 1995 and ceased operation in April 1998 following exhaustion of its liquid helium supply. It contained an infrared telescope of 60 cm aperture with several different focal-plane instruments, as detailed below. The details of the instrumentation have been given in several publications. A convenient source of information concerning the satellite, its performance and the calibration of the instruments, available in the astronomical literature, is the November special issue of *Astronomy and Astrophysics*, 1996, **315**, L27–L400.

2.8.1.1 ISOPHOT – imaging photometer/polarimeter

This covered 2.5–240 μm; could be used for multi-aperture photometry, polarimetry, imaging and spectrophotometry.

2.8.1.2 ISOCAM – camera

Offered two 32 × 32 detector arrays, an InSb array for the shorter wavelengths and a Ga-doped Si photoconductive array for the longer. Bandpass filters and circular variable filters (CVFs) were available for each. Each pixel was 1.5, 3, 6 or 12 arcsec across. Polarimetry was also an option.

2.8.1.3 SWS – short-wavelength spectrometer

The coverage was 2.4–45 μm with spectral resolution of 1000–2000 in normal grating mode. From 11.4–44.5 μm the resolution was increased 20 times by means of a Fabry–Perot etalon. Depending on the wavelength, different apertures and detector types were used.

2.8.1.4 LWS – long-wavelength spectrometer

This instrument covered the spectral range 43–197 μm with resolution of $\sim$170 to $\sim$8000, divided at $\lambda = 70\,\mu$m into two ranges. The high-resolution mode used a Fabry–Perot etalon.

2.8.2 NICMOS camera on HST

The Near Infrared Camera and Multi-Object Spectrometer on the Hubble Space Telescope (HST) is designed to cover the spectral range 0.8–2.5 μm (Axon et al., 1996). The detectors are three 256 × 256 HgCdTe arrays. Each detector is attached to an independent camera, the fields being 51.2 × 51.2, 19.2 × 19.2 and 11 × 11 arcsec^2. Each camera has a filter wheel which contains a variety of filters, polarizers or grisms. One camera has a coronagraphic occulting spot. The resolution of the grisms is 200.

The main advantage of the NICMOS camera is freedom from the absorption, emission and seeing effects caused by the terrestrial atmosphere. Longwards of $1.8\,\mu$m, thermal emission from the HST's optics dominates the background from the zodiacal dust.

2.8.3 Future space observatories

2.8.3.1 WIRE

The Wide-field Infrared Explorer (WIRE) is a 28 cm cryogenically-cooled telescope which will make a survey of limited portions of the sky at 12 and $25\,\mu$m to a depth about 1000 times fainter than IRAS. Its main function will be to investigate the luminosity function of starburst galaxies at high redshift and to study their evolution. Status information is available at http://sunland.gsfc.nasa.gov/smex/wire.

2.8.3.2 SIRTF

The Space Infrared Telescope Facility (SIRTF) is a NASA project due for launch in 2001. Passive radiation to space will assist with the cooling. It will have a 0.85 m cryogenic telescope with imaging and spectroscopic capabilities from 3–$180\,\mu$m. Its lifetime will be 2.5 yr. Its principal instruments are: (1) Infrared Array Camera (IRC) – simultaneous imaging at 3.5, 4.5, 6.3 and $8\,\mu$m with 256×256 arrays; (2) Infrared Spectrograph (IRS) with resolution $= 50$ or 600 in 5–$40\,\mu$m range; and (3) Multiband Imaging Photometer (MIPS) – simultaneous imaging at 12, 30, 70 and $160\,\mu$m, with a spectrometer covering 50–$100\,\mu$m at a resolution of 20. Information is available at http://sirtf.jpl.nasa.gov/sirtf/home.html.

2.8.3.3 IRIS

The IRIS (Infrared Imaging Surveyor) is a 70 cm cryogenic telescope being prepared by the Institute of Space and Astronautical Science (Japan) for launch in 2002. It will cover the spectral range 2–$200\,\mu$m. A 1 yr lifetime is expected, with extensions at the short-wavelength end of its range. The telescope will be maintained for the first year at <7 K. The far-IR detectors will be cooled to <2 K.

The principal instruments will be (1) the Near/Mid Infrared Camera (IRC), with large-format InSb and Si : As arrays, a filter wheel and a grism spectrometer; and (2) the Far-Infrared Surveyor (FIS) covering

50–200 μm with Ge : Ga and stressed Ge : Ga arrays. The FIS will have a Fourier spectrometer (R $\sim$ 200) (Shibai and Murakami, 1996).

2.9 Infrared databases

Two comprehensive printed databases are:

Gezari et al. (1993a), listing infrared photometry and spectroscopy and including a bibliography of the infrared astronomical literature. Sources are ordered by their position in the sky.

A sub-section of the above, concerning observations at $\lambda > 4.6\,\mu$m, is given in Gezari et al. (1993b).

Beichman, C. A., Neugebauer, G., Habing, H. J., Clegg, P. E. and Chester, T. J., 1988. *The Infrared Astronomical Satellite (IRAS): Catalogs and Atlas*, Vols. 1–7, NASA RP-1190, Washington, D.C. (The definitive results from IRAS).

World Wide Web:

A useful source of IRAS maps and other information concerning past and future US infrared experiments is the Infrared Processing and Analysis Center's (IPAC) World Wide Web site http://www.ipac.caltech.edu.

Simbad is a useful general site which includes many infrared data and cross-identifications; http://simbad.u-strasbg.fr.

2.10 Further reading

Allen, D. A., 1975. *Infrared, the New Astronomy*, Shaldon, Devon, UK. Anecdotal account of IR astronomy in 1975.

Elston, R., (ed.), 1990. *Astrophysics with Infrared Arrays*, ASP Conf. Ser., 14, Astr. Soc. Pacific, San Franscisco.

Epchtein, N., Omont, A., Burton, B. and Persi, P., (eds.) *Science with Astronomical Near-Infrared Surveys*, Kluwer, Dordrecht. A useful book about the ground-based infrared surveys and what is expected from them.

Signore, M. and Dupraz, C., 1991. *The Infrared and Submillimeter Sky after COBE*, Kluwer, 1991.

3

Photometry

3.1 Infrared photometry

3.1.1 General

Ground-based infrared photometry divides naturally at about $5\,\mu$m into two wavelength regimes: *JHKLM* at the shorter wavelengths and *NQ* at the longer. The first of these corresponds to the sensitivity range of InSb detectors. The *JHKLM* filters have effective wavelengths $1.25\,\mu$m, $1.65\,\mu$m, $2.2\,\mu$m, $3.5\,\mu$m and $4.8\,\mu$m. Longer wavelengths are covered by bolometers and various photoconductive or photovoltaic detectors cooled to temperatures <10 K. The commonest filters are those mentioned: $N(10\,\mu$m) and $Q(20\,\mu$m), though many variants and sub-bands have been used.

Each of *JHKLMNQ* corresponds to a more or less clear "window" of atmospheric transmission (see Figs. 3.1–3.5). The traditional *L*-band is often replaced by the more optimal L', centred at $3.8\,\mu$m. The *M* and *Q* are the least satisfactory, being interrupted by numerous features due to water-vapor absorption.

Infrared *magnitudes* are based on the usual logarithmic scale where

$$m_\lambda = -2.5 \log R + ZP$$

where R is the instrumental response in the relevant band, assumed linear, and ZP is a constant called the *zero-point*.

The *absolute magnitude* of an object is its apparent magnitude if it were placed at a distance from us of $10\,\mathrm{pc}$ (3.086×10^{17} m).

3.1.2 Zero-points

In photometry, objects are measured relative to standard stars, and the absolute level is a quantity which must be set in some way. A common procedure in the northern hemisphere is to define the bright A0V star Vega as having zero magnitude at all wavelengths and to relate all other stars to this.

It is also possible to base the infrared zero-points on the pre-existing visible system so that the average of $(V-m_\lambda)$ for all A0V stars is zero for all bands, whatever their wavelength λ.

Johnson et al. (1966), using the second approach, determined the magnitude of Vega at *JHK* to be 0.02 at each wavelength. It is a difficult star to observe with high accuracy from the southern observatories because of its large minimum zenith angle. It has also been found to be one of the main-sequence stars which possess a circumstellar dust shell affecting its measured magnitudes beyond $20\,\mu$m, making it unsuitable for use as a standard at longer wavelengths. For these reasons, southern observers have preferred to relate their zero-points to the visible system via A0V stars.

3.1.3 Effective wavelengths

The *effective wavelength* of a filter is defined as

$$\lambda_{\text{eff}} = \frac{\int \lambda S(\lambda)\eta(\lambda)\,\mathrm{d}\lambda}{\int S(\lambda)\eta(\lambda)\,\mathrm{d}\lambda}$$

where $S(\lambda)$ is the transmission of the filter and $\eta(\lambda)$ is the quantum efficiency of the detector. In some infrared wavebands, the transmission of the Earth's atmosphere plays an important part in the definition of the band.

3.1.4 Derivation of monochromatic flux densities from photometry

The calibration of a standard star system is normally given as the flux corresponding to a zeroth magnitude star at the effective wavelength of each band. Knowing by how many magnitudes a measured object is fainter than the standard, its flux may be calculated. However, the derivation of the correct value is not really as simple as this, except when the filter bandwidth is narrow. More care is necessary in the case of broad bands. The task may be even be impossible, as when the spectrum is

Table 3.1. *Factor K calculated for various blackbodies*

Temperature	K_J	K_H	K_K	K_L	K_N
5000	0.99	1.00	1.00	1.00	0.99
3000	0.96	0.98	0.99	0.99	0.99
1000	1.03	0.98	0.98	0.98	0.95
800	1.14	1.01	0.98	0.97	0.94
600	1.42	1.09	1.00	0.97	0.92
400	2.62	1.44	1.09	0.99	0.90
300	5.48	2.10	1.25	1.04	0.92

Notes: The standard star is assumed to have a 10000 K blackbody spectral energy distribution. The detectors were modelled as a photon detector with responsivity proportional to λ for *JHKL* and a bolometer (responsivity independent of λ) for *N*. The *JHKL* filters were taken to be those of the SAAO (Carter, 1990) system with a hypothetical *N*-band from 8 to 14 μm. The precise values of the *K*-factors will differ from one photometric system to another.

composed of emission lines. Even when the monochromatic flux density $F_\lambda(\lambda)$ varies smoothly with wavelength, a non-linear rise across the band may occur, such that the flux at the effective wavelength would be overestimated if it is derived simply from the magnitude. This problem is particularly acute for cool blackbody distributions.

Because of the difficulties associated with conversion of measurements into absolutely calibrated monochromatic fluxes, photometrists often find it convenient not to take this step at all but choose instead to discuss their results in terms of colors and magnitudes.

Correction tables can nevertheless be constructed for various assumed intrinsic spectral energy distributions and particular filter transmission curves. The correction factor K, such that

$$F_\lambda^{\text{true}}(\lambda_{\text{eff}}) = F_\lambda^{\text{apparent}}(\lambda_{\text{eff}})/K$$

for a particular monochromatic flux density distribution $F_\lambda(\lambda)$, can be calculated as follows:

$$K = \frac{\int F_\lambda(\lambda)S(\lambda)\,d\lambda}{F_\lambda(\lambda_{\text{eff}})} \bigg/ \frac{\int F_\lambda^{\text{std}}(\lambda)S(\lambda)\,d\lambda}{F_\lambda^{\text{std}}(\lambda_{\text{eff}})} ,$$

where the integration is taken over the filter band of interest and $S(\lambda)$ is a factor containing the transmission of the atmosphere and optics and the response of the detector. $F_\lambda^{\text{std}}(\lambda)$ is the flux density distribution of the standard star. This calculation has to be carried out for the particular set of equipment and atmospheric conditions relevant to the observations being corrected. Some examples (Table 3.1) are given for a set of particular circumstances. Note that the corrections are very minor for blackbodies covering the effective temperature range of normal stars. A similar calculation shows that power-law distributions need corrections of only a few percent for exponents $-4 < \alpha < 0$, where $F_\lambda \propto \lambda^\alpha$ (or $2 > \alpha > -2$ if $F_\nu \propto \nu^\alpha$).

Another example was computed by Low and Rieke (1974) for blackbody distributions and their $HKLMNQ$ filter set. In the particular case discussed it was assumed that the standard stars have effective temperatures of around 4000 K.

Flux correction factors are also discussed in connection with photometry obtained from the IRAS satellite in section 3.6.1.

3.1.5 Isophotal wavelengths

In the previous subsection it was shown that the monochromatic flux at the effective wavelength of a broad-band filter can be corrected for a particular spectral distribution. An alternative approach, encountered in calibration work, is to find the wavelength at which the simply-derived value for F_λ is the correct one.

The *isophotal wavelength* of a filter and star combination is defined to be that quantity λ_i which satisfies the relation

$$F_\lambda(\lambda_i) \int S(\lambda)\, d\lambda = \int F_\lambda S(\lambda)\, d\lambda = R$$

where $F_\lambda(\lambda)$ is the monochromatic flux density from the star in question, say in units of W m^{-2} sec^{-1} μm^{-1}, and $S(\lambda)$ is the efficiency of the photometric system (Golay, 1974). R is proportional to the response of the system (the output voltage of the detector, for example).

3.1.6 Photometric systems

A *photometric system* is usually defined in practice by presenting the magnitudes of a set of standard stars measured according to that system. Other objects can be referred to it by using the same photometer that was used to measure the standards. However, this is not practicable if more than one observatory wishes to refer measurements to the same standard system. The main problem that arises between different observatories is that the interference filters that define the passbands cannot be reproduced with perfect accuracy. Some degree of uniformity can be obtained by having all filters made in the same batch manufacturing process and by observing the same set of standard stars.

It is often necessary in real life to make comparisons with results obtained from different equipment. The best way to do this is to measure the defining standards with the new filters and determine formulae (known as *color equations*) through which the systems can be related. These formulae are imperfect, because they break down when extrapolated beyond the color range of the standard stars and also when they are applied to objects with strong emission lines or deep absorption bands in their spectra.

3.2 Infrared photometric bands

3.2.1 JHKLM photometry

The first published photometry in these bands was by Johnson (1962) and it covered K (2.2 μm) and L (3.6 μm). Later on Johnson (1964) published a set of J (1.20 μm), K (2.20 μm), L (3.5 μm) and M (5.0 μm) photometry for 256 stars, which became the defining set of standards for the Arizona *JKLM* system. At first, and until the appearance of photovoltaic InSb (indium antimonide) detectors (*ca* 1974), the *J(H)KL* observations were made with chemically-deposited PbS (lead sulphide) cells, whose sensitivity cut off (at 77 K) around 3.8 μm. The *H*-band (1.65 μm) was introduced and used systematically only somewhat later (e.g., Johnson, MacArthur and Mitchell 1968). With the general introduction of photovoltaic InSb detectors, some observatories decided to switch from the original *L*-band to L'(3.8 μm).

Variations of the *K*-band, designated K' and K_S, are now also used. These have a shorter cut-off wavelength in order to reduce the contribution of the terrestrial thermal background radiation, which begins to be significant in this part of the spectrum (Wainscoat and Cowie, 1992). The passband of K_S is from 2.0 to 2.3 μm while that of K' is from 1.9 to 2.3 μm. The short cut-on wavelength of the K' filter is intended to take advantage of the reduced water-vapour extinction above Mauna Kea.

While the *J*- and *H*-bands are not affected by ambient thermal radiation, the sensitivities in the K, L and M are degraded by it.

3.2.1.1 JHKLM filters and the atmosphere

The filters used at *JHKLM* are designed to match the clear or almost clear bands in the transmission of the atmosphere. The most significant

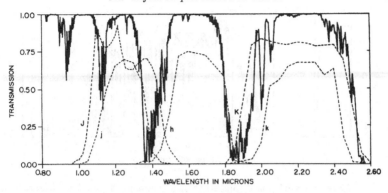

Fig. 3.1. The transmission from 0.80 to 2.60 μm of the atmosphere above Kitt Peak in summertime, from Manduca and Bell (1979). The Johnson (1965) J and K "sensitivity functions" and the Kitt Peak J, H and K filter transmissions (marked jhk) are also shown.

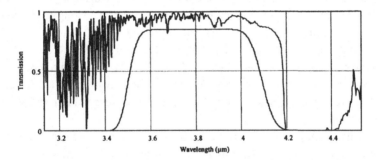

Fig. 3.2. L' region with atmospheric transmission modelled for Mauna Kea and an idealized filter curve (Simons, 1996).

sources of atmospheric opacity are water vapour and carbon dioxide. The water-vapor content, in particular, varies greatly with latitude, altitude, temperature and time of year. The best (lowest) extinctions are obtained at the high-altitude sites such as Mauna Kea, Hawaii (4200 m), during the coldest weather, but high-altitude sites near the South Pole should prove even better.

Figure 3.1 (Manduca and Bell, 1979) shows a theoretical atmospheric transmission curve for Kitt Peak (a site representative in altitude (2080 m) and latitude of many active observatories) in summertime. On it are superimposed Johnson's (1965) "response functions" and transmission curves for the Kitt Peak National Observatory's JHK filters, which are typical of those in current use.

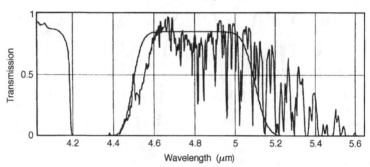

Fig. 3.3. *M* region with atmospheric transmission modelled for Mauna Kea and a typical (idealized) filter curve (Simons, 1996).

Looking at Fig. 3.1, it is clear that the original Johnson J and K filters are undesirably broad and that variations in the atmospheric water-vapor content will affect their effective wavelengths. The long-term trend has been towards somewhat narrower passbands with sharper cut-offs so as to avoid this problem. Nevertheless, L and L' (Fig. 3.2) are still affected by absorption bands below $3.5\,\mu$m. The M-band (Fig. 3.3) (usually centered at $4.8\,\mu$m in current practice) is heavily affected at all times by deep absorption lines so that the extinction is always high and somewhat uncertain.

Simons (1996) considers the specification of new near-infrared filter bandpasses suited to conditions above Mauna Kea, Hawaii, and at sites with altitude $\sim 2\,$km for the Gemini telescopes.

Young, Milone and Stagg (1994) are amongst those who have advocated narrower bandpasses than those currently in use. In their view, the edges of the filters should be less steep in order to facilitate the transformation of data between systems having (almost inevitably) slightly different filter specifications.

3.2.2 Narrow-band CO and H_2O photometry

Considerable photometric work has been done using sub-bands of the K-band to isolate regions affected by CO and water vapor. For example, Persson, Aaronson and Frogel (1977) discuss the narrow-band properties of late-type dwarf stars. This work was also extended to galaxies (see section 3.9 for references). Water vapor was measured with a narrow-band filter centred at $2.00\,\mu$m and CO with another at $2.36\,\mu$m.

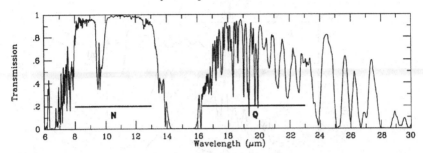

Fig. 3.4. Atmospheric transmission from 6–30 μm for Mauna Kea. The air-mass was taken to be 1.15 and the precipitable water-vapor content 1 mm in this calculation by Lord (1992). The ranges of typical wide-band N and Q filters are shown as horizontal bars. Note the deep dip at 9.6 μm caused by O_3 high in the atmosphere. Figure is from Tokunaga (1999).

A third narrow-band filter around 2.20 μm was used to define the un-affected continuum. Comparisons between results obtained by different groups are not always straightforward because they are very sensitive to the precise effective wavelengths of the filters, which are not always identical.

3.2.3 Mid-infrared, ground-based photometry

Ground-based aperture photometry in the mid-infrared bands has re-ceived little attention in recent years, largely because of the difficulties involved in detecting faint or extended objects in the presence of very high backgrounds. The IRAS satellite, in particular, has produced a large database of observations at longer wavelengths which are supe-rior for many purposes to what can be done from the ground. However, the availability of mid-infrared cameras on large telescopes has led to a recent revival of interest.

Most of the mid-infrared photometric systems described in the litera-ture relied on Ge-doped bolometers, but contemporary cameras mainly use various types of doped Si arrays as their detectors (see section 6.5.7).

3.2.3.1 The N-band region (8–14 μm)

A band of fairly good atmospheric transmission covers from 8 to 14 μm with a dip at about 9.6 μm due to ozone absorption high in the atmo-sphere (Fig. 3.4). The designation N is used for filters which cover most

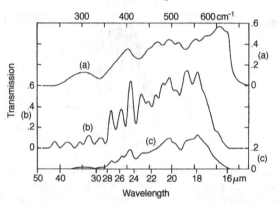

Fig. 3.5. Q-band region: (a) Transmission of cooled $20\,\mu$m filter, (b) atmospheric transmission for $1\,$mm of H_2O and (c) product of (a) and (b); from Simon, Morrison and Cruikshank (1972).

of this range. Few filter transmission curves have been published. An "N" filter supplied to SAAO in 1982 consisted of a long-pass interference filter with a sharp cut-on around $7.7\,\mu$m, a transmission of about 80% to $\sim$14 μm and a long-wave blocker of BaF_2, $1\,$mm thick, which cuts off transmission at about $15\,\mu$m. The long-wavelength cutoff in this system is actually provided by the earth's atmosphere and the BaF_2 blocker merely serves to reduce the thermal background seen by the detector. The system used by Rieke, Lebofsky and Low (1985) for setting up their photometric system at 10 and $20\,\mu$m, on which the IRAS calibration is based, covered from about 9.8 to $12.6\,\mu$m (half-maximum points), thus avoiding the uncertainty due to the absorption bands at the boundaries of the region.

The N-band region is sometimes divided into narrower bands. For example, Thomas, Hyland and Robinson (1973) used bands centered at $8.4\,\mu$m and $11.2\,\mu$m besides the broad-band N, which they took to be at $10.2\,\mu$m. At ESO, there are three narrow bands, $N1$, $N2$ and $N3$, centered at 8.4, 9.69 and $12.9\,\mu$m respectively. Low and Rieke (1974) mention the use of bands at 11.5 and $13\,\mu$m. Young, Milone and Stagg (1994) discuss the possible creation of two narrow bands, at 9.0 and $11.0\,\mu$m, which should be clear of telluric features.

3.2.3.2 The Q-band region (17–27 μm)

The Q-band is often poorly defined, being affected by water-vapor absorption over its whole width of about $10\,\mu$m, stretching from 17 to $27\,\mu$m, according to atmospheric water-vapour content. Simon, Morrison and Cruikshank (1972) show typical filter and atmospheric transmission curves (See Fig. 3.5). Rieke, Lebofsky and Low (1985) give a normalized reponse curve for their system.

3.3 Standard star observations

Standard stars should ideally be numerous so as to be available in any part of the sky where objects are to be measured. Apart from the general requirement of non-variability and high precision, they should also cover a wide range of normal spectral types so that objects can be compared to standards having similar colors. The latter is not always very practical in the infrared, where highly reddened objects are frequently observed. The complex dependence of photometric accuracy on stellar color and seasonal changes in the Earth's atmosphere are discussed by Manduca and Bell (1979).

There should also be a wide range of magnitudes covered by the system, so that linearity of detector response can be judged from them, and so that they can be used on both small and large telescopes as well as with array detectors, which usually saturate at comparatively low light levels.

Observational programs for the establishment of standard star systems have become uncommon since the advent of infrared array detectors, which are relatively difficult to use for precision work. Many observatories have discontinued the use of photometers and, furthermore, the work is time-consuming and tedious. Nevertheless, standards remain essential and it is necessary to understand the limitations of those currently available.

In the SAAO standard program of Carter (1990), the fundamental standards were chosen as far as possible to be in or near the Harvard E-regions – a group of nine areas equally spaced around the sky at −45° declination. They were intercompared frequently at strictly similar air masses, east and west of the meridian. Adjacent regions were compared as well as every second region. In the course of a year, the circle of comparisons could be joined. The small "closing error" was distributed amongst the comparisons. The ultimate accuracy of the comparisons was probably limited by minute flexure effects in the photometer-telescope combination as well as non-uniformity of response as the star is moved across the aperture of the photometer.

The program also included other stars at a constant air mass equal to that of the E-regions in order to provide for general sky coverage. It was particularly important to include equatorial stars so that northern hemisphere observers could link to the system if they wished. Also, standard stars from other observatories' programs were observed in order to derive the color equations between systems.

Particular emphasis was also given to observing dwarf stars around A0 spectral type so that the zero-points of the infrared colors $V-K$, $J-H$, $H-K$ and $K-L$ could be set to zero for dwarfs with $B-V = 0.0$. This is a more reliable procedure than setting the zero-point from the average of a number of A0V stars, since the reddening vectors in the color-color diagrams are very nearly parallel to the locus of color vs. spectral type.

Some other systems, such as that advocated by Bessell and Brett (1988) have had their zero-points set so that Vega would have color zero in all indices.

3.3.1 JHKL standard star programs – particular

3.3.1.1 Johnson's (1964) JKL system – the original

Johnson's (1964) original standard work was of a very high quality for the time and was regarded as adequate for at least a decade. Unfortunately, it did not extend very far into the southern hemisphere. Southern observers have since filled this gap, such as Glass (1974) in Cape Town and Carter (1990) at Sutherland (SAAO), Engels et al. (1981) and Bouchet, Manfroid and Schmider (1991) at La Silla (ESO), Jones and Hyland (1982) at Mount Stromlo, Allen and Cragg (1983) at the Anglo-Australian Observatory (AAO), Elias et al. (1982, 1983) at Caltech and Cerro Tololo and McGregor (1994) at Mount Stromlo (MSSSO).

3.3.1.2 SAAO (Carter) system

The SAAO standards list, as measured by Carter, has been extended to cover a group of fainter stars in a cooperative program with the AAO (Carter and Meadows, 1995), particularly to cater for infrared array cameras.

As a result of the high internal consistency of the Carter program, it became worthwhile to intercompare the standard star systems of all the different southern observatories. This was done by Glass (1985), Carter (1990), Bessell and Brett (1988) and McGregor (1994). The result of this was that color equations have been established with fair accuracy, and the general quality of the systems can be judged. The internal consistency of the Carter and CTIO–Caltech systems was found to be the highest, although no judgment could be made as to which was the better. It is evident that accuracies better than 0.01 mag are achieved by both. McGregor (1994) shows that the new MSSSO standards also possess high internal accuracy.

It is clear that when photometry of the highest practicable accuracy is undertaken, great care must be taken to ensure that the measurements are made on a standard system with traceable filter combinations. The use of ill-defined sets of filters will hamper comparisons with other work and also lead to uncertainty in the application of reddening corrections, which depend on the precise characteristics of the filters.

3.3.1.3 The SAAO (Glass) system of 1974

The Glass (1974) system was a first attempt to provide a set of reliable standards to extend the Johnson set for observers south of the equator and to provide H standards. Its transformation properties to other systems are well known (Glass, 1985; Carter, 1990) and the existence in it of small R.A.-dependent terms (amounting to 0.04 mag at worst) has been discussed by Carter (1990). It has been replaced at SAAO by the Carter (1990) system.

3.3.1.4 MSO system

The current Mount Stromlo system is based on a J filter from the same batches as the SAAO (Glass) one, but the HKL ones are slightly different. However, McGregor (1994) regards the system as essentially the same as the Bessell and Brett one. He gives transformations to the CTIO, the original MSO system (Jones and Hyland, 1982), the AAO, CTIO–Caltech and SAAO systems. He also gives a graph of the relative quantum efficiency of the detector (Cincinnati) vs wavelength, the filter transmissions and a set of standard stars.

3.3.1.5 ESO standard systems

The original ESO system set up by Engels et al. (1981) had a rather peculiar H filter which led to marked color differences when compared to most of the other systems in use (Glass, 1985; Carter, 1990). In addition, its random errors appeared to be relatively high. The later ESO system, based on filters in use since 1981, is defined by the 199 standard stars in Bouchet, Manfroid and Schmider (1991). Its filters are more conventional and the measurements are of much better internal consistency. Bouchet et al. give transformations enabling conversions to be made to several other systems.

Besides JHK, the ESO system includes the L'- and M-bands.

A study by Bersanelli, Bouchet and Falomo (1991), comparing the
ESO standards with various other sets, indicates that there are differences which affect the magnitudes but not the colors. These have the
effect of limiting their magnitude accuracy to ± 0.03 mag and their color
accuracy to ±0.01 mag.

3.3.1.6 Bessell and Brett's system

The system of Bessell and Brett (1988) is essentially a rationalized version of the Glass (1974) system and that of Johnson (1964). It incorporates data from various authors and discusses the transformations to
the other main systems. It effectively replaces the homogenized system
of Koorneef (1983a, 1983b). The L' and M filters are also discussed.

3.3.1.7 CTIO–Caltech system

The Elias et al. (1982, 1983) lists include the Caltech standards. The
J filter of this system has a relatively long effective wavelength, so that
comparison with other systems requires a considerable color term.

3.3.1.8 AAO system

This set of standards was produced by Allen and Cragg (1983). It comp-
rises over 60 stars, many in common with other systems. Attempts were
made to include stars with a wide range of colors in order to facilitate
transformations between systems.

3.3.1.9 UKIRT and related standards

A set of faint standards ($8 < K < 14$ mag) on the natural system of
the UKIRT telescope, together with transformation equations to the
CTIO–Caltech system, have been given by Casali and Hawarden (1992).
The coverage of this system has been extended in numbers and range
of spectral types by Hunt et al. (1998), who used similar filters to the
UKIRT ones.

3.3.1.10 Persson et al. (1998) faint standards

These comprise a grid of 65 faint ($10 < K < 12$) stars spread around the
sky with the object of being useful for array work. The J-, H-, K- and
K_s-bands were covered. Transformation equations to the UKIRT and
CTIO–Caltech systems are provided as well as filter transmission curves.
The data were obtained with NICMOS arrays. The method of reduction
and the precision of the results are discussed in detail.

3.3.2 Transformations between systems

It is important to be able to compare measurements between systems. Simple relations known as *color equations*, which relate measurements of different systems, can usually be found for stars with well-behaved spectra, i.e., those that do not have deep absorption bands or prominent emission lines. A number of these have been tabulated by Bessell and Brett (1988), Glass (1985) and Carter (1990).

3.3.3 Standard stars for the mid-infrared bands

Standard star systems for mid-infrared ground-based photometry have been set up by Simon et al. (1972) and Tokunaga (1984) at Q. Thomas, Hyland and Robinson (1973) and Tokunaga (1984) review previous work and discuss the absolute calibration of the system. Rieke, Lebofsky and Low (1985) give a list of 10 standards for N and Q in connection with their calibration work which forms the basis for the absolute calibration of the IRAS photometry. The spectrophotometric work of Cohen and collaborators (see section 4.5.1) is a more recent potential source of mid-IR photometric standards.

Because the N- and Q-bands are so wide, particular care is necessary in extracting monochromatic flux densities from photometry (see section 3.1.4).

3.4 Colors of normal stars

It is important that the behaviour of normal stars be known if peculiarities due to such factors as membership of different populations are to be studied effectively. Colors for stars in the various luminosity classes and most normal spectral types were derived by Johnson (1966) for his $UBVRIJKL$ system. Lists of intrinsic infrared colors ($VIJHKLL'M$) on their homogenized Johnson–Glass system were published by Bessell and Brett (1988), for dwarfs and giants (class III); see Tables 3.2 and 3.3. These authors give color equations for transformation to other systems.

Figures 3.6 and 3.7 show the colors of solar-neighbourhood giant (classes II and III) stars as actually measured on the Carter system, taken from standard star and other work (Glass, 1997). Quite good agreement with the Bessell and Brett (1988) tables is found after applying the appropriate transformations.

There are quite considerable variations of intrinsic color, depending on population group, among late-type stars. For example, M giants in

Table 3.2. *Intrinsic colors of dwarfs on the Bessell and Brett (1988) system*

MK	V–K	J–H	H–K	J–K	K–L	K–L'	K–M
B8	−0.35	−0.05	−0.035	−0.09	−0.03	−0.04	−0.05
A0	0.00	0.00	0.00	0.00	0.00	0.00	0.00
A2	0.14	0.02	0.005	0.02	0.01	0.01	0.01
A5	0.38	0.06	0.015	0.08	0.02	0.02	0.03
A7	0.50	0.09	0.025	0.11	0.03	0.03	0.03
F0	0.70	0.13	0.03	0.16	0.03	0.03	0.03
F2	0.82	0.165	0.035	0.19	0.03	0.03	0.03
F5	1.10	0.23	0.04	0.27	0.04	0.04	0.02
F7	1.32	0.285	0.045	0.34	0.04	0.04	0.02
G0	1.41	0.305	0.05	0.36	0.05	0.05	0.01
G2	1.46	0.32	0.052	0.37	0.05	0.05	0.01
G4	1.53	0.33	0.055	0.385	0.05	0.05	0.01
G6	1.64	0.37	0.06	0.43	0.05	0.05	0.00
K0	1.96	0.45	0.075	0.53	0.06	0.06	−0.01
K2	2.22	0.50	0.09	0.59	0.07	0.07	−0.02
K4	2.63	0.58	0.105	0.68	0.09	0.10	−0.04
K5	2.85	0.61	0.11	0.72	0.10	0.11	
K7	3.16	0.66	0.13	0.79	0.11	0.13	
M0	3.65	0.695	0.165	0.86	0.14	0.17	
M1	3.87	0.68	0.20	0.87	0.15	0.21	
M2	4.11	0.665	0.21	0.87	0.16	0.23	
M3	4.65	0.62	0.25	0.87	0.20	0.32	
M4	5.26	0.60	0.275	0.88	0.23	0.37	
M5	6.12	0.62	0.32	0.94	0.29	0.42	
M6	7.30	0.66	0.37	1.03	0.36	(0.48)	

Taken from Bessell and Brett (1988).

the Galactic Bulge show less deviation from the blackbody curve than those in the solar neighbourhood (Frogel and Whitford, 1987). Similar effects are apparent in M giants observed far from the Plane, towards the Galactic poles (Feast, Whitelock and Carter, 1990).

M dwarf colors also show strong population dependence (Leggett, 1992). There is a progression from higher *(J–H)* color to lower as we go from young disc to halo populations.

3.4.1 Influence of molecular absorption on broad-band colors

The deviation of giant stars in the *J–H*, *H–K* diagram from the locus of blackbodies, reaching a maximum at around M2–M3 for giants, is attributable to the minimum of the continuous bound-free and free-free

Table 3.3. *Intrinsic colors of giants on the Bessell and Brett (1988) system*

MK	V–K	J–H	H–K	J–K	K–L	K–L'	K–M
G0	1.75	0.37	0.065	0.45	0.04	0.05	0.0
G4	2.05	0.47	0.08	0.55	0.05	0.06	−0.01
G6	2.15	0.50	0.085	0.58	0.06	0.07	−0.02
G8	2.16	0.50	0.085	0.58	0.06	0.07	−0.02
K0	2.31	0.54	0.095	0.63	0.07	0.08	−0.03
K1	2.50	0.58	0.10	0.68	0.08	0.09	−0.04
K2	2.70	0.63	0.115	0.74	0.09	0.10	−0.05
K3	3.00	0.68	0.14	0.82	0.10	0.12	−0.06
K4	3.26	0.73	0.15	0.88	0.11	0.14	−0.07
K5	3.60	0.79	0.165	0.95	0.12	0.16	−0.08
M0	3.85	0.83	0.19	1.01	0.12	0.17	−0.09
M1	4.05	0.85	0.205	1.05	0.13	(0.17)	−0.10
M2	4.30	0.87	0.215	1.08	0.15	(0.19)	−0.12
M3	4.64	0.90	0.235	1.13	0.17	(0.20)	−0.13
M4	5.10	0.93	0.245	1.17	0.18	(0.21)	−0.14
M5	5.96	0.95	0.285	1.23	(0.20)	(0.22)	−0.15
M6	6.84	0.96	0.30	1.26			0.0:
M7	7.8	0.96	0.31	1.27			0.0:

Taken from Bessell and Brett (1988).

absorption due to H$^-$ ions. Lee (1970) found that the deviation was greater for luminosity class III than for more and less luminous spectral classes. This unexpected result, at least for the more luminous stars, was explained theoretically by Bell et al. (1976) as being due to strengthened line opacity of CO and CN in the supergiants.

At temperatures below about 3250 K, (corresponding to spectral type M6) the effects of water vapor in the atmospheres begin to dominate those of other absorbers.

The *M*-band at around 4.8 µm is strongly affected by CO absorption so that the *L–M* colors of giants are zero or negative at late M spectral type.

Bessell et al. (1989a) have made extensive model atmosphere calculations for M giant stars, showing the sensitivity of various color indices including the broad-band infrared colors to gravity, temperature, atmospheric extension and metallicity. A more recent paper (Bessell, Castelli and Plez, 1998) reports progress on the derivation of broad-band colors from updated model atmospheres and discusses their relationship to experimental data.

Photometry

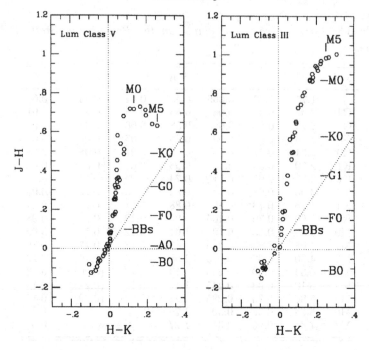

Fig. 3.6. Measured *J–H* and *H–K* colors of normal dwarf (left) and giant (right) stars on the SAAO (Carter) system, for the solar neighbourhood; compiled from Carter's (1990) standards, Carter and Meadows (1995), Whittet and van Breda (1980, transformed, additional early-type stars), Feast, Whitelock and Carter (1990, additional M-type giants) and Mould and Hyland (1976, transformed, additional late-type dwarfs). BBs denote the loci of blackbodies.

The dependence of *J–K* color on metallicity is predicted to be quite strong; typically $\delta(J–K) \sim 0.15$ mag for $\delta\,(Z/Z_\odot) = 1.5$, in the sense that *J–K* becomes bluer for low metallicities. The effect is slightly less for the *J–L* color. In the two-color *J–H, H–K* diagram, the deviation of the giant models from the blackbody line increases for low metallicities. Effects of this kind have been observed in globular clusters (Frogel et al., 1990). There is also a less prominent tendency for giants to be further from the blackbody line than supergiants, as first noted by Lee (1970).

The early- to solar-type dwarfs occupy about the same parts of the two-color diagrams as the giants. However, as K and M types are approached, water vapor again dominates over the H⁻ opacity (Jones et al., 1994), and their colors become more blackbody like, though well separated from those of the giants.

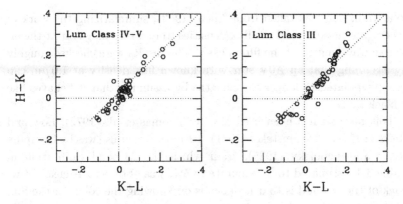

Fig. 3.7. Measured *H–K* and *K–L* colors of normal dwarf (left) and giant (right) stars on the SAAO (Carter) system, for the solar neighbourhood; compiled from Carter's (1990) standards, Whittet and van Breda (1980, transformed, additional early-type stars), Feast, Whitelock and Carter (1990, additional M-type giants) and Mould and Hyland (1976, transformed, additional late-type dwarfs). The loci of blackbodies are shown as dotted lines.

Mira variables are cooler and more luminous than earlier-type M giants. They show prominent water-vapour bands in their spectra and their location in the *J–H, H–K* diagram shows them to be closer to the blackbody line.

3.5 Absolute calibration

3.5.1 *Absolute calibration at JHKLM*

Three approaches have been used to obtain absolute calibration of *JHKLM* photometry, although the most direct and fundamental method – that of directly comparing a calibrated laboratory blackbody standard to a star – has been somewhat neglected by recent authors.

3.5.1.1 *Solar method*

The first absolute calibration of *JKL* (together with some other wavebands) was obtained by Johnson (1965). Firstly, the *V–J, V–K* and *V–L* colors of the Sun (of G2V spectral type) were estimated from observations of several stars believed to be solar analogues. Secondly, the apparent *V* mag of the Sun was taken to be –26.74, the weighted mean of six independent previous determinations, and its apparent *JKL* magnitudes were derived. Finally, the absolute energy distribution of the solar radiation was taken from Allen (1963), summarizing the work of others, and used to find the flux densities from a zero mag star at the effective wavelengths of the filters used. The results were checked roughly by assuming that an A0V star with known flux density at $1.0\,\mu$m can be extrapolated to longer wavelengths by assuming that it behaves like a black body.

This method has been used also by Thomas et al. (1973), Low and Rieke (1974) and Wamsteker (1981). Its most recent use was by Campins, Rieke and Lebofsky (1985), from which the M-band calibration in Table 3.4, estimated to be accurate to 5%, was obtained. A clear drawback of the method is that it depends on knowing the colors of the Sun in the standard bands and these are by no means easy to determine. It also requires knowledge of both the Sun's absolute energy distribution and its apparent visual magnitude.

3.5.1.2 Blackbody comparison method

Two main problems arise in the direct comparison of stellar fluxes with terrestrial blackbody sources. Firstly, the signal from the blackbody is likely to be many orders of magnitude greater than that from a star and, secondly, the effect of atmospheric absorption over a broad band is difficult to determine with accuracy.

The first comparison using this method was made by Walker (1969) using moderately broad-band filters centered at 1.06, 1.13, 1.63 and $2.21\,\mu$m. The detector was a PbS cell cooled to dry ice temperature. A 402 K blackbody was mounted above the photometer. Although geometrical and reflection losses were not precisely known, the overall accuracy of the procedure was estimated at 10%.

A series of measurements based on narrower bands of good transmission, around 2.20 and $3.76\,\mu$m, were obtained by Selby et al. (1983). A final accuracy of $\pm4\%$ was claimed. Further measurements in several bands were reported by Blackwell et al. (1983).

3.5.1.3 Comparison with stellar models

Photometry may also be calibrated by relying on theoretical models of bright stars for extrapolation of measured visible-region fluxes to infrared wavelengths. A recent paper using this scheme is that of Cohen et al. (1992a) which takes models of Vega and Sirius by Kurucz, normalized to measurements at $5556\,\text{Å}$, and uses them to calculate isophotal wavelengths and monochromatic flux densities for various commonly-used photometric systems. The $JHKL$ calibration given in

Table 3.4. *Absolute calibration of photometry – flux densities for a star with mag* $= 0$

Band	λ_{eff} μm	ν_0 Hz	F_λ $\mathrm{W\,cm^{-2}\,\mu m^{-1}}$	F_ν $\mathrm{W\,m^{-2}\,Hz^{-1}}$	$\log F_\nu$	$\log \nu_0$
U	0.366	8.19×10^{14}	4.175×10^{-12}	1.790×10^{-23}	-22.75	14.913
B	0.438	6.84×10^{14}	6.32×10^{-12}	4.063×10^{-23}	-22.39	14.835
V	0.545	5.50×10^{14}	3.631×10^{-12}	3.636×10^{-23}	-22.44	14.740
R_C	0.641	4.68×10^{14}	2.177×10^{-12}	3.064×10^{-23}	-22.51	14.670
I_C	0.798	3.79×10^{14}	1.126×10^{-12}	2.416×10^{-23}	-22.62	14.575
J	1.22	2.46×10^{14}	3.15×10^{-13}	1.59×10^{-23}	-22.80	14.390
H	1.63	1.84×10^{14}	1.14×10^{-13}	1.02×10^{-23}	-23.01	14.26
K	2.19	1.37×10^{14}	3.96×10^{-14}	6.4×10^{-24}	-23.21	14.14
L	3.45	8.7×10^{13}	7.1×10^{-15}	2.9×10^{-24}	-23.55	13.94
M	4.8	6.3×10^{13}	2.2×10^{-15}	1.70×10^{-24}	-23.77	13.80
N	10.6	2.8×10^{13}	9.6×10^{-17}	3.60×10^{-25}	-24.44	13.55
Q	21	1.43×10^{13}	6.4×10^{-18}	9.4×10^{-26}	-25.03	13.15

Note: Sources for $UBVR_C I_C JHKL$ are Bessell, Castelli and Plez (1998). Note that $V = 0.03$ for Vega on this system. The $JHKL$ colors are on the Bessell and Brett system; M, Campins, Rieke and Lebofsky (1985); NQ, Rieke, Lebofsky and Low (1985).

Table 3.4 is taken from the recent work of Bessell, Castelli and Plez (1998).

3.5.2 Calibration of mid-infrared photometry

As in the case of near-infrared photometry, the calibration methods used in the mid-infrared can be regarded as direct or indirect. The latter depend on model atmosphere calculations.

In the direct methods, a bright object is used as an intermediate between a laboratory standard and the standard infrared stars. Rieke, Lebofsky and Low (1985) discuss three separate direct calibrations. One of these is essentially the solar method, described in section 3.5.1.1, and the other two make use of Mars as an intermediate standard. The magnitude of Mars was obtained by comparison with an Earth-bound blackbody, using a specially stable bolometer detector and compensating for differences in the atmospheric transmission. It was also measured with a previously calibrated radiometer aboard a space probe. Mars could then be compared with a network of ten carefully intercompared standard stars, which included Vega. The estimated accuracy of the calibration of Vega was given by Rieke et al. (1985) at about ±3%. The results by the direct method were consistently about 10% higher than those from the indirect (see also the discussion of more recent model atmospheres in section 3.8.1).

The direct calibration of the Q-band via Mars is also discussed by Rieke et al. (1985). The absolute calibration of the Mars photometry was obtained only by the calibrated space probe radiometer. A much higher uncertainty (±8%) is associated with this measurement because of the difficulties caused by uncertain terrestrial atmospheric absorption.

The Rieke et al. (1985) photometric system and calibration, quoted in Table 3.4, is based on $N = Q = 0.02$ for Vega.

A discussion by Cohen et al. (1992a) of the indirect or stellar atmosphere method of calibration yields 35.9 Jy at 10.5 µm for $N = 0$, if we take Vega to have $N = 0.02$. This compares very closely with the Rieke et al. value of 36.0 ± 1.2 Jy at 10.6 µm.

3.6 IRAS photometry

The IRAS (Infrared Astronomical Satellite) has made the greatest contribution to long-wavelength infrared photometry.

The 12, 25, 60 and 100 µm fluxes or upper limits for more than 245,000 objects have been tabulated, covering almost the whole sky and including (in many cases) indications as to variability in the *IRAS Catalogs* (Beichman et al., 1988). The *Explanatory Supplement* to the *IRAS Catalogs* is the main source of detailed information for interpretation of the Catalog entries.

The absolute calibration of IRAS photometry made use of the 10.6 µm absolute ground-based measurement of α Tau by Rieke, Lebofsky and Low (1985) and extrapolation of its flux to 25 and 60 µm. The 100 µm calibration was obtained by extrapolating the fluxes from three asteroids that had been measured at 25 and 60 µm.

Various color-color diagrams based on IRAS photometry, sometimes in conjunction with other wavebands, have been devised for diagnostic purposes. Given here (Fig. 3.8) is an example of a $[12] - [25]$ vs. $[25] - [60]$ diagram from van der Veen and Habing (1988).

3.6.1 IRAS color corrections

The IRAS filters are very wide ($\Delta\lambda/\lambda \leq 0.7$) and the response of the system changes considerably with wavelength. Considering that the spectral

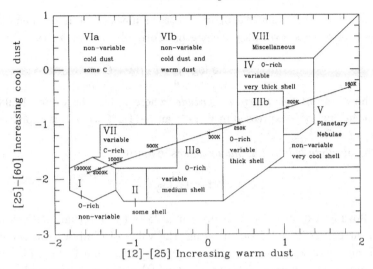

Fig. 3.8. Example of an IRAS two-color diagram from van der Veen and Habing (1988). This diagram separates mass-losing late-type stars such as Miras and OH/IR stars into various approximate categories according to the temperatures of their dust shells. The loci of blackbodies are also shown. The colors are computed from the IRAS fluxes as $2.5 \log_{10}(F_{25}/F_{12})$, etc.

energy distributions of the objects measured vary widely, the quoted flux will be appropriate to some isophotal wavelength, unique to each object, which may not be close to the nominal or effective wavelength of each band (see sections 3.1.3–3.1.5).

The fluxes in the Catalog are derived on the assumption that the objects have spectral energy distributions such that $\nu F_\nu = \lambda F_\lambda = \text{constant}$.

The *Explanatory Supplement* shows how the quoted fluxes may be corrected to give the true values at the effective wavelengths λ_0 of 12, 25, 60 and 100 μm or the effective frequencies ν_0 of 25, 12, 5 and 3 $\times$ 10^{12} Hz of each band, when some color and spectral shape information is available. Tables of correction factors for the fluxes at each effective wavelength are given (p. VI-26). The catalogued colors may be used to determine the appropriate power-law spectral index or blackbody temperature, and hence the appropriate corrections.

3.7 Bolometric magnitudes

Theoretical models of stars usually predict their luminosities and effective temperatures. In this section and the following, the derivation of these quantities from observations is considered.

The *bolometric magnitude* of an object is a measure of its total energy output, integrated over all wavelengths. Formally, we can write

$$m_{\text{bol}} = -2.5 \log \int_0^\infty F_\lambda \, d\lambda + C$$

for the *apparent bolometric magnitude*, where C has the value –18.980 when the energy is expressed in $W\,m^{-2}$ and –11.480 when it is expressed in $\text{erg}\,\text{cm}^{-2}\,\text{s}^{-1}$.

The *absolute bolometric magnitude* is given by

$$M_{\text{bol}} = 4.74 - 2.5 \log \left(\frac{L}{L_\odot} \right).$$

Following a discussion at Commission 25 of the International Astronomical Union in its Kyoto meeting, the value of C above was set by *defining* the absolute bolometric magnitude of the Sun to be 4.74, corresponding to a measured *solar constant* (the average integrated solar flux received at the earth) of $1360\,W\,m^{-2}$.[†] The need for an arbitrary definition of the absolute solar bolometric magnitude arises mainly from a lack of agreement as to which set of standard stars should be used when setting the minimum absolute value of the bolometric correction BC_V (see section 3.8 for definition). $BC_{V,\text{min}}$ occurs near spectral type F5 on the main sequence and is traditionally intended to be zero (R. Cayrel, communication to members of Commission 25).

Ideally, the bolometric energy is determined by integrating a detailed spectral energy distribution, but this is rarely available due to the limitations of atmospheric transparency and other practical problems. Moderately high accuracies can be obtained for the cooler stars from broadband photometry by interpolation, extrapolation to zero flux at the limits and modelling the overall distribution by a blackbody curve. However, very cool stars with deep absorption bands cannot be treated reliably in this way.

3.8 Stellar effective temperatures

The effective temperature T_{eff} of a star is derived from the relation

$$F = \left(\frac{\phi}{2} \right)^2 \sigma T_{\text{eff}}^4$$

where F is the total observed flux, ϕ is its observed angular diameter

[†] Corresponding to $L_\odot = (3.826 \pm 0.008) \times 10^{26}$ W.

and σ is Stefan's constant. The value of ϕ may be obtained from lunar occultations or Michelson stellar interferometry (see Fig. 3.9) for suitably close (resolvable) objects. It must be corrected for limb-darkening effects.

Ridgway et al. (1980) have compiled a list of effective temperatures from measured angular diameters of K0–M6 giants with suitable photometry. They give a table of spectral type vs T_{eff}, $V-K$ and $I(104) - L$ colors.

The effective temperatures of Mira variables, which are also giants, vary by several hundred degrees around their cycles. The atmospheres of these stars are very tenuous and the normal definitions of effective radius and effective temperatures no longer apply. Essentially, there is no clearly defined "surface," or radial distance from the center of the star, at which the optical depth of the atmosphere increases sharply for all wavelengths. A fairly satisfactory approach is to define the *effective radius* as that where the radial Rosseland mean optical depth is 1. Model atmospheres have been calculated by Bessell et al. (1989b), though the difficulties involved are quite severe.

Di Benedetto (1993) discusses empirical effective temperatures and angular diameters, obtained by Michelson stellar interferometry, for G–K stars. He also gives tables of bolometric corrections and effective temperatures for stars of $1.42 \leq (V-K)_0 \leq 7.60$ and luminosity classes I–V.

The *bolometric correction* BC_X is given by

$$m_{bol} = m_X + BC_X$$

where the X refers to a particular passband. Tabulations of temperature vs. $V-K$, $J-K$ and other colors for the appropriate luminosity and metallicity, based on the most recent theoretical models, can be used to obtain the bolometric corrections BC_V or BC_K (Bessell, Castelli and Plez, 1998).

3.8.1 Infrared flux method

The infrared flux method for obtaining effective temperatures, introduced by Blackwell and Shallis (1977), depends on the belief that the prediction by theoretical modelling of the infrared flux from a star is likely to be accurate because line-blanketing effects are less than in the visible and stellar opacities are better understood at long wavelengths. A measurement of a star's bolometric magnitude and a single near-infrared continuum point should thus be enough to determine its T_{eff}.

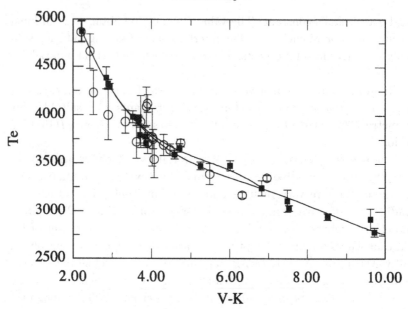

Fig. 3.9. Empirical T_{eff} vs. $V-K$ diagram for giants. Occultation data from Ridgway et al. (1980) are shown as open circles. Michelson stellar interferometer data are shown as closed squares. Error bars in the measured temperatures are shown. The long line is the current recommended fit, while the shorter one refers to the Ridgway et al. (1980) temperature scale; taken from Bessell, Castelli and Plez (1998).

In a discussion of the limitations of the infrared flux method, Mégessier (1994) shows that the the measured ratio

$$R = \sigma T^4 / F_\lambda$$

is sensitive to metallicity effects and gravity, besides temperature. The metallic lines effectively reduce the UV flux, causing more energy to emerge in the infrared, for a given effective temperature. The effect on the apparent value of T_{eff} is of order 1%, while different gravities may change it by half as much. The results may still be model-dependent, there being an effect on the derived T_{eff} of order 2% between the oldest and most recent Kurucz (ATLAS) codes.

3.9 *JHKL* photometry of galaxies

What might be termed "ordinary" galaxies, i.e., members of the Hubble sequence excluding irregulars and "active" galaxies, have surprisingly uniform colors.

Table 3.5. *Infrared colors of "ordinary" galaxies as a function of type*

Classes	T Types	No. in Sample	$J-K$	s.e. of Mean	log A/D (0) (mean)
Ellipt.		13	1.00	0.01	−0.67
Sa, Sab	1,2	6	1.03	0.01	−0.59
Sb, Sbc	3,4	12	1.01	0.02	−0.78
Sc, Scd	5,6	5	0.97	0.03	−0.92
Sd, Sdm	7,8	2	0.95	0.04	−1.05
Sm	10	1	0.78	0.07	

Note: This is taken from Glass (1984).

Surface photometry of a number of S0 and spiral galaxies has been given by Terndrup et al. (1994), with photometry at rJK, where r is a variant of Cousins' R band at $0.64\,\mu$m. Although small color gradients often exist in $r-K$ and $J-K$, there is little systematic trend with Hubble type.

The near-IR colors of disc and bulge components are very similar, indicating that they resemble each other in their contents of late-type stars.

Table 3.5 gives average colors of ordinary nearby galaxies as a function of position on the Hubble sequence (early to late type galaxies) from Glass (1984), on the SAAO (Carter) system. Color gradients were not detected in this work, but their apparent absence may have been due to a rather limited range of log A/D(0) values (ratio of aperture size to apparent galaxy size). Color indices involving a visual band and an IR one, such as $V-K$, do, however, become markedly redder for later types.

Extensive data on the visible to near-infrared colors, including CO and H_2O indicies, of normal galaxies are available from the work of Frogel, Persson, Aaronson and Matthews (1978), Aaronson, Frogel and Persson (1978) and Persson, Frogel and Aaronson (1979).

3.9.1 Relativistic K-corrections

Photometry of moderately redshifted objects can be related to source quantities in an approximate way by the use of K-corrections, which are first-order adjustments that compensate for the changes of wavelength, bandwidth and intensity associated with redshift. Note that "K" is a generic term for these corrections and does not apply to the K filter only.

Near-IR observations offer the advantage that the corrections, especially at K, are generally much smaller than in the UBV bands.

Table 3.6. *Typical K-corrections*

Color	Corrections
J–K	$-0.5z$
H–K	$-3.5z$
K	$3.3z$

Note: These quantities must be added to the observed quantity to get the rest-frame quantity; from Frogel et al. (1978).

K-corrections can be calculated if the spectrum of the galaxy is known or can be predicted with reasonable accuracy. Frogel et al. (1978), for example, used balloon observations of the spectra of several late-type stars as templates (Table 3.6). The derived corrections are approximately linear up to $z = v/c = 0.2$. Comprehensive tables of K-corrections have been computed by Poggianti (1997).

The general formula for the K-correction in a particular filter i is as follows:

$$K_i = -2.5 \log(1 + z) + 2.5 \log \left(\frac{\int_0^\infty F_\lambda(\lambda) S(\lambda)\, d\lambda}{\int_0^\infty F_\lambda \left(\frac{\lambda}{(1+z)} \right) S(\lambda)\, d\lambda} \right),$$

where $S(\lambda)$ is the response of the filter and photometric system, z is the redshift and $F(\lambda)$ is the observed flux density from the galaxy. The first term in the expression arises from the narrowing of the filter passband in the rest-frame of the galaxy by a factor $(1 + z)$ and the second term allows for the fact that radiation seen by the observer at wavelength λ is emitted at wavelength $\lambda/(1 + z)$.

It is useful to remember that a power-law spectral distribution undergoes Lorentz transformation without change of spectral index, and that a blackbody spectral distribution transforms to another blackbody distribution, but with lower temperature. The observed temperature is given by

$$T_{\text{obs}} = T(1 - \beta \cos \theta)(1 - \beta^2)^{-1/2}$$

where θ is the direction of the motion to the line of sight and β has the usual meaning of v/c.

If details of the spectral distribution are unknown, or cannot be reasonably predicted, the K-corrections cannot be derived. For example, quasar colors can be affected strongly when emission lines of substantial equivalent width fall within the photometric passbands.

3.9.2 Photometric determination of redshifts

Because of the general difficulty of obtaining spectra of faint galaxies, considerable attention has been paid to obtaining redshifts by use of multicolor photometry. For example, Connolly et al. (1995) find that redshifts can be determined within ±0.05 for galaxies out to $z \sim 0.8$ and $B_J < 22.5$, using filters similar to the standard $UBRI$ bands, irrespective of galaxy type and luminosity. The method depends principally on the effect that the 4000 Å break in galaxy spectra has on their redshifted colors, but ceases to be useful at about $z = 1$, where this feature passes beyond the I-band. It again works well for $Z > 2.2$, where the Lyman limit (912 Å) enters the U-band.

The problem of determining z for intermediate redshifts $1 \leq z \leq 2$ has been addressed by Connolly et al. (1997), simply by including the J-band within their scheme. They have applied their method to determining how the general star formation rate depends on z (see also section 5.12.2.1).

3.9.3 Effects of star formation

When star formation is taking place, the visual-region and near-infrared colors of a galaxy may be made bluer due to the presence of many hot, young, stars. An early model for the time dependence of galaxy colors, caused by continuous star formation or burst-like events, was constructed by Struck-Marcell and Tinsley (1978). Glass and Moorwood (1985) interpreted a number of starburst galaxies in terms of their positions in the $U - B$, $V - K$ diagram and Thuan (1983) gives colors for a sample of blue compact dwarf galaxies (BCDGs).

3.9.4 Modelling galaxy Evolution

The evolution of galaxies or clusters can be modelled by assuming that stellar populations are generated, either initially or continuously, according to certain given conditions, such as metallicity and mass function. Stellar evolution theory is then applied to each type of star to follow its development. The overall spectrum of the galaxy at any given time is calculated by adding together all the individual contributions. The

expected broad-band photometric colors and other quantities can then
be derived. Additional features of the model can include the interstellar
medium and its interaction with the evolving stars. The accomplish-
ment of this task requires extensive libraries of stellar and other spectral
energy distributions. A suitable compilation is available on CD-ROM
and is described by Leitherer et al. (1996), together with a review of
evolutionary models.

3.9.5 Galaxy colors at high redshift

The observed colors of galaxies at high redshifts will be functions of
both evolution and redshift. Poggianti (1997) considers both effects in
a model for normal galaxies up to $z = 3$ and shows that the changes to
be expected are much smaller in the near-infrared than in, for example,
the U-, B- and V-bands. Considering also that the effect of interstellar
absorption is also reduced at longer wavelengths, the near-infrared thus
has a clear advantage for studies of the luminosity functions of galaxies. According to Mobasher, Sharples and Ellis (1993), E/S0 and spiral
galaxies have identical luminosity functions, within the errors, at K.
The slope of the relationship between K magnitude and log (number of
galaxies $\deg^{-2}$ mag^{-1}) begins to turn over at $K \sim 19$.

Preliminary examination of the Hubble Deep Field and other deep
surveys appears to show that star formation reached a maximum at a
time corresponding to a redshift of about $z = 2$ (see Ellis, 1997, for a
review). Since the dust content of galaxies (see Chapter 5) is dependent
on the availability of heavy elements produced through star formation,
their far-infared behavior will also depend on their ages and evolutionary
history.

3.10 Suggestions for further reading

Golay M., 1974. *Introduction to Astronomical Photometry*, Reidel, Dordrecht and Boston.   Though mainly concerned with visible region
photometry, many topics of general interest in the theory of precision
photometry are covered.

Milone E. F., (ed.)   1989.   *Infrared Extinction and Standardization*,
Springer-Verlag, Berlin. Mainly concerned with *JHKLM* photometry.

Sterken C. and Manfroid J., 1992. *Astronomical Photometry, A Guide*,
Kluwer, Dordrecht. Deals with the issues of visible-region photometry,
many of which are also of concern in the infrared.

4

Spectroscopy

4.1 Introduction

We can consider spectroscopy at several different resolution levels.

The crudest resolution amounts to photometry, where the wavelength scale is divided into a small number of regions, or wide bands, to give an indication of a *spectral energy distribution* (SED). Examples might include the $U(0.37)$, $B(0.44)$, $V(0.55)$, $R(0.64)$ and $I(0.80\,\mu\text{m})$ system in the visible region and the $J(1.25)$, $H(1.65)$, $K(2.2)$, $L(3.5)$, $M(4.8)$, $N(10)$, and $Q(20\,\mu\text{m})$ system of ground-based infrared, as well as the 12, 25, 60 and $100\,\mu\text{m}$ IRAS bands.

Many data with $\Delta\lambda/\lambda \sim 0.01$ exist from spectrometers that used circular variable filters (CVFs). This is usually considered to be the least resolution which can really be called spectroscopy.

The next level might be called medium-resolution spectroscopy, where the detailed lines that usually compose molecular bands are not seen, but the bands are treated in an averaged manner. Most infrared spectroscopy falls into this category.

With high-resolution spectroscopy the individual molecular lines can be examined. Sensitivity considerations have limited this type of work hitherto to the study of the Sun, bright stars and planets. With the advent of large infrared array detectors and large telescopes, it is reasonable to predict that fainter objects will soon receive attention. High resolution has in the past been obtained through the use of Fabry–Perot etalons or Michelson interferometers, also called Fourier–transform spectrometers.

4.2 Stellar spectra

The theoretical study of stars is usually divided into *stellar interiors* and *stellar atmospheres.* The output of the interior can be regarded as something like a blackbody, but the atmosphere modifies the emergent flux according to its temperature structure, pressure structure and composition.

High-gravity stars such as dwarfs (luminosity class V) have atmospheres which are thin compared to their radii, but low-gravity stars such as giants and supergiants have tenuous atmospheres which contain relatively little mass and can occupy a very large part of the "volume" of a star.

What is normally observed is the *photosphere*, which is usually well-defined in a high-gravity star, but which can have a rather vague meaning in the case of a giant or supergiant. Its outer boundary is often defined by the radius at which the star has a certain optical depth, for example, $\tau_\lambda = 1$. The stellar diameter thus becomes a function of wavelength. The diameter of a Mira is 50% or more greater in strong absorption lines than in the $2\,\mu$m continuum (Bessell et al., 1989a).

Outside the photosphere can be a chromosphere which usually radiates only a very small part of the total or *bolometric* energy of the star. Also, a star can be surrounded by a dust shell, created from the medium out of which it formed or by mass loss from the star itself. Such a shell absorbs energy in the form of short-wavelength photons from the star and re-radiates it at infrared wavelengths.

4.3 Opacity sources in the infrared

The absorption in a stellar atmosphere arises from various physical processes and may be continuous or piecewise continuous across large wavelength ranges or may only affect particular wavelengths or bands.

4.3.1 Continuous opacity sources

In the infrared part of the spectrum of hot stars, i.e., those hotter than A0 ($T_{\text{eff}} = 9400\,$K for Vega; Cohen et al., 1992a), free-free transitions of hydrogen and helium provide sources of opacity which vary smoothly with wavelength, and the emergent spectra resemble blackbodies, with features due to the various absorption lines of H and He.

In moderately hot stars ($>7000\,\mathrm{K}$), continuous absorption is dominated by neutral hydrogen. Hotter stars (O and B types) are dominated by He.

The cooler stars, such as the Sun (G2V) and later types, are dominated in the visible and infrared by the H^- ion.

The continuous mean opacities in a zero-metallicity gas, i.e., one containing only hydrogen, its molecules, and helium, have been calculated by Lenzuni, Chernoff and Salpeter (1991), taking into account collision-induced absorptions, bound-free absorption and Rayleigh and Thomson scattering.

4.3.1.1 H^- bound-free

Stars cooler than A0 are dominated by H^- bound-free and free-free absorption. The ionization potential of H^- is 0.75 eV, corresponding to a photon wavelength of 1.655 μm. The bound-free absorption coefficient of H^- reaches a maximum at about 0.84 μm, while the free-free absorption coefficient rises continuously towards the infrared. The combined absorption from these sources is lowest near 1.6 μm. For this reason, the spectra of late-type stars have a broad "bump" around the same wavelength.

A detailed discussion of the H^- absorption coefficient is given by Gray (1976). Bell et al. (1976) discuss the effect of luminosity on the bump arising from the H^- opacity minimum in K to early M stars.

4.3.2 Atomic and molecular opacity

4.3.2.1 Atomic lines

Kleinmann and Hall (1986) have identified a number of atomic absorption lines in the 2.0–2.5 μm region in late-type stars of various luminosities. These include NaI, FeI, MgI, CaI, TiI, SiI and AlI. They discuss the strength of Brγ, CaI and NaI lines as a function of T_{eff}. This work was extended to dwarf stars by Ali et al. (1995), including also MgI, and expressing the line strengths as equivalent widths. The latter authors state that the temperature can be determined from equivalent widths to an accuracy of about $\pm350\,\mathrm{K}$, assuming that the metallicity is solar. Ramírez et al. (1997) have found, however, that the Na and Ca absorption features in cool, luminous, late-type stars in the Galactic Center cluster are contaminated by ScI and other elements. Using some new high-resolution observations, they show that an index depending on the equivalent widths of CO, Na and Ca can be useful in the two-dimensional (temperature and luminosity) classification of obscured late-type dwarfs and giants.

4.3.3 Molecular bands

A recent discussion of the effects of CO and H_2O absorption in the 1.6–2.5 μm region of late-type stars and their use as population diagnostics has been given by Lançon and Rocca-Volmerange (1992), based on observed spectra.

4.3.3.1 CO

The ratio of O to C in a cool stellar atmosphere is of critical importance in determining its overall spectrum. The high dissociation energy of the CO molecule ensures that all the C and O atoms will combine to form CO until nothing is left of one or the other. When uncombined O is left over, the star will be of M type, dominated by oxygen compounds such as TiO, SiO and H_2O. When C remains, the star will be of C type, showing molecules such as C_2, CN, CH, HCN and C_2H_2. When O and C are about equal, so that both are bound up as CO, less abundant molecules will dominate the spectrum, giving rise to S-type stars. CO is thus ubiquitous in late-type stellar atmospheres.

The CO fundamental vibration ($\nu = 1 - 0$) band around 4.7 μm, the ($\nu = 2 - 0$ and $\nu = 3 - 1$) bands around 2.3 μm, and the ($\nu = 3 - 0$ and $\nu = 4 - 1$, etc.) bands around 1.65 μm affect K and later types quite strongly. Detailed synthetic spectra in the region of the 2.3 μm bands have been computed by Bell and Briley (1991) for various dwarf and giant models at a resolution of ~1600 and have also been used to generate narrow-band photometric indices. CO starts to become detectable in dwarfs at temperatures around 6500 K.

CO ($\nu = 2 - 0$) emission is seen in some stellar types such as young stellar objects (YSOs), hot supergiants and post-AGB stars (see, e.g., Oudmaijer et al., 1995).

4.3.3.2 H_2O

In very cool stars (late M types), water-vapor absorption dominates the near-infrared spectra.

The spectra of late M dwarfs resemble the wavelength dependence of the absorption coefficient of water vapor (Jones et al., 1994). From the assumption that the water-vapor bands dominate the near-infrared spectra of dwarfs, these authors have derived effective temperatures and radii which can be compared with theoretical models.

4.3.3.3 SiO

A feature due to the fundamental of SiO has its maximum at $\sim$7.9 μm. This can be observed best from high-altitude airplanes or space. The first overtone is at $\sim$4.1 μm. The increasing strength of the fundamental band with spectral type has been described by Cohen et al. (1995). Its effects become noticeable at G8III and increase to K5III, continuing at about the same level to late M types.

4.3.3.4 H_2: Collision-induced absorption

The inability of the H_2 molecule to radiate by dipole processes is relieved under conditions of high pressure when collisions between H_2 and H_2, H_2 and He, H_2 and H set up "virtual supermolecules" which possess dipole moments and can interact with photons with much greater probability than normally (Lenzuni, Chernoff and Salpeter, 1991). These processes are important in the high-gravity atmospheres of red dwarfs and brown dwarfs, and can be the dominant source of atmospheric opacity when their metallicities are low.

The absorption takes place in wide bands centered around the normal rotation and vibration-rotation frequencies and make significant contributions to the opacity between 0.7 and 40 μm. Simultaneous transitions can occur in both colliding partners so that absorption and emission can take place at the sum and difference frequencies of the individual species. The collision times are very short ($\Delta t \sim 10^{-12}$), so that the bands are broad, typically 100 cm^{-1} at T = 3000 K.

4.4 Model atmospheres

A recent review of cool star model atmospheres is given by Jørgensen (1997). Bessell, Castelli and Plez (1998) discuss the agreement of colors derived from Plez's NMARCS and other recent codes with observational data.

4.4.1 G–K dwarfs

Bell and Gustafsson (1989) gave calculated effective temperatures for G and K dwarf stars of various gravities and abundances. Synthetic visible and near-infrared colors were tabulated. Bell and Briley (1991) gave a grid of models at a resolution of ~3000 for both cool giants and dwarfs in the region of the 2.3 μm CO bands. In addition to comparing the spectra with the observations of Kleinmann and Hall (1986), narrow-band CO indices were calculated for comparison with photometric observations. The luminosity sensitivity of the CO band strength was discussed by Bell and Tripicco (1991), who found that there is a lesser tendency for C to be tied up in CO in giants than in dwarfs.

4.4.2 M dwarfs

Late-type dwarfs have been particularly difficult to model, mainly because of uncertainty in the precise absorption due to water vapor. Allard and Hauschildt (1995) give a grid of model atmospheres and discuss the opacities. They clearly show the transition from metal rich types to metal poor types, in which the dominant absorber at $\lambda > 1.5\,\mu$m goes from H_2O to H_2–H_2 collision-induced opacity. This explains the disappearance of the H^- opacity minimum at 1.6 μm and the marked change with metallicity which occurs in the *J–H, H–K* diagram of dwarfs.

The fit of the best available synthetic spectra to observed spectra is still unsatisfactory, mainly in the 1.6 μm band, although improved knowledge of the water molecular lines has led to better fits at longer wavelengths. This problem is discussed by Allard et al. (1994). The opacities determined experimentally by Ludwig (1971) appear to be too high.

Tsuji, Ohnaka and Aoki (1996) suggest that the condensation of dust in the atmosphere can be very important for $T_{\mathrm{eff}} < 2800\,$K and its inclusion in models can improve the fit to real spectra by decreasing the depth of the molecular absorption. The effective temperatures are deduced to be much lower than those in the Allard and Hauschildt (1995) models.

Model atmospheres of cool, low-metallicity stars are treated by Borysov, Jørgensen and Zheng (1998), who include collision-induced absorption for H_2–H_2 and H_2–He and also use complete line data for the molecules TiO, H_2O, CN, CH and SiO.

4.4.3 G–K giants

Spectra of G–K giants around 2.3 μm were also computed by Bell and others referred to in section 4.4.1 on G–K dwarfs.

4.4.4 M giants

Scholz (1985) and Bessell et al. (1989a,b) consider the extended atmospheres of M giants, including Miras. They consider the behavior of photometric colors as gravity, temperature, extension and metallicity are varied.

The density and temperature structures of Mira atmospheres is computed. The monochromatic radii ($\tau_\lambda = 1$) are calculated for maximum and minimum states of the variation in sample stars. The effects of the H^- opacity minimum and the strong bands of CO and H_2O on the spectra are shown.

4.4.5 Brown dwarfs and giant planets

Brown dwarfs are low-mass "stellar" objects similar to giant planets, differing from them in their external circumstances and possibly in the manner of their formation. Brown dwarfs may also be distinguished from giant planets by having experienced a period of deuterium fusion reactions. Neither type of object supports hydrogen fusion.

The near-IR spectrum of the brown dwarf Gl 229B, which has a temperature of $\sim$900 K, is dominated by deep bands of CH_4 and H_2O (Geballe et al., 1996). Its methane absorption bands are somewhat similar to those observed to dominate the spectra of Uranus, Neptune and the Saturnian satellite, Titan (Fink and Larson, 1979; Geballe et al., 1996). The strength of the absorption in the K-band in Uranus is so great that its rings become more conspicuous than the planet itself.

Model atmospheres of very-low-mass stars and brown dwarfs have been reviewed by Allard et al. (1997). Burrows et al. (1997) present model atmospheres which extend down to 100 K and point out that the near-infrared fluxes of cool objects may be enhanced by several orders of magnitude over the expectation from a blackbody. They also discuss the probability of detecting extrasolar planets and brown dwarfs with some current and future satellite and ground-based observatories.

4.4.6 Cool carbon stars

The spectra of cool carbon stars in the 1.5–2.4 μm region are very different from those of late M-type stars. As a result of the dominance of CO, as discussed in section 4.3.3.1, small differences in relative C and O abundance affect the atmosphere quite dramatically. C_2, CN and CO are prominent in carbon-star spectra (Lambert et al., 1986).

Spectra from 2.4 to $40\,\mu$m and R = 300 − 1000 have been obtained by Yamamura et al. (1998); see Fig. 5.3. The most prominent features at this resolution are absorption bands around 3.1 and $5.0\,\mu$m. High-resolution spectra around $3\,\mu$m by Ridgway, Carbon and Hall (1978) show many lines arising from HCN and C_2H_2.

At longer wavelengths there is a SiC dust feature at $11.2\,\mu$m and another feature at $30\,\mu$m. At around $13.7\,\mu$m there is an absorption band due to the C–H bending mode.

4.5 Infrared spectral atlases

Table 4.1 contains a list of papers dealing with specific classes of stars at low, medium and high resolutions.

4.5.1 Spectrophotometric calibration

Spectrophotometric observations depend on (ideally) featureless stars near the objects of interest which may be used to remove spurious absorption lines of telluric (caused by the Earth's atmosphere) origin by division. Bright and truly featureless stars are not available, so an alternative has to be found. Early-type O stars can be regarded as blackbodies in the infrared, except for absorption lines of helium at 1.572 and $1.692\,\mu$m. Origlia, Moorwood and Oliva (1993) show how to remove the effects of these features.

Nearby dwarf stars of late F and early G spectral types may also be used to compensate for atmospheric absorption, even though they contain numerous absorption lines. The spectrum of the object to be corrected is divided by that of a suitable late F or early G dwarf at similar airmass and the result is multiplied by the solar spectrum to remove artifacts caused by absorption features in the dwarf spectrum. Maiolino, Rieke and Rieke (1996) discuss the corrections, such as wavelength adjustments and resolution changes, that must be applied to a standard solar spectrum in order to use this technique.

A series of absolutely calibrated spectrophotometric standard stars has been constructed by M. Cohen (Table 4.2) and his collaborators (see Cohen in the list of references). They depend on models of Vega and Sirius due to Kurucz, normalized to the visual flux densities of these stars at 5556 Å (Cohen et al., 1992a). An extension was made to cover α Tau from 1 to $35\,\mu$m (Cohen, Walker and Witteborn, 1992b). This work and similar methods were used for the absolute calibration of the ISO SWS (short-wavelength spectrometer), which covers 2.4–45 μm, by Schaeidt et al. (1996).

Table 4.1. *Spectral atlasses in the infrared*

Spectral Region	Spectral Types	Reference
Low resolution		
1.2–5.5 μm[1] (1.5%)	KIII–MIII	Strecker et al. (1979)
1–2.5 μm (150–300)	late MV	Tinney et al. (1993)
Medium resolution		
2.15–2.35 μm (1380)	F3V–M6V	Ali et al. (1995)
1.15–2.42 μm (500)	YSOs; MKV stds	Greene and Lada (1996)
2–2.2 (2.4) μm (1000)	O, B, Oe, Be	Hanson et al. (1996)
1–2.5 μm (300)	M2V–M9V	Jones et al. (1994)
2–2.5 μm (3000)	F8–M7, I–V	Kleinmann and Hall (1986)
1.43–2.5 μm (500)	O6–M8.5, I–V	Lançon and Rocca (1992)
1.0–2.5 μm (500)	Ae, Be, I	McGregor et al. (1988)
1.45–2.4 μm (1000)	Of, WN, Be, LBV	Morris et al. (1996)
1.53–1.72;	AOI–M5I;	Origlia et al. (1993)
	G8III–M7III;	*" " " "*
2.29-2.3 μm (1500)	G2V, K1V	*" " " "*
1.6–2.4 μm (400–700)	Post-AGB	Oudmaijer et al. (1995)
2.19–2.34 μm (1380, 4830)	KOIII–M6III	Ramírez et al. (1997)
3.98–4.07 μm (450)	KOIII–M7III, others	Rinsland and Wing (1982)
1.44–2.5 μm (2500)	N-type	Tanaka et al. (1990)
2–2.4 μm (3000)	O–M, I–V	Wallace and Hinkle (1997)
High resolution		
0.9–5.3 μm (10^5)	K2III	Hinkle et al. (1995)
1.5–2.5 μm ($\leq 10^5$) and 3.6–4.2 μm	Cool C stars	Lambert et al. (1986)
0.7–22 μm (3×10^5)	Sun	Wallace et al. (1996)
2.02–2.41 μm ($\geq$45,000)	F–M, I–V; C, S	Wallace and Hinkle (1996)

1. Spectrophotometry.

The LWS (long-wavelength spectrometer) of ISO covers the spectral range 43–197 μm. Its initial calibration depends on the spectrum of Uranus, as measured by Hanel et al. (1986) to 50 μm, using a spectrometer on board Voyager, together with a semi-theoretical model extending to the longer wavelengths (Swinyard et al., 1996).

4.6 Interstellar medium

The interstellar medium is characterized according to its density n (number of atoms per cm^3). To the basic classification scheme as given by Whittet (1992) we may add photodissociation regions and HII regions which are excited by stars.

Table 4.2. *Spectrophotometric standards in the infrared*

Star	Spectral Types
Vega $= \alpha$ Lyr	A0V
Sirius $= \alpha$ CMa	A1V
α^1 Cen	G2V
α TrA	K2III
ϵ Car	K3III
α Boo	K1III
γ Dra	K5III
α Cet	M1.5III
γ Cru	M3.3III
μ UMa	M0III
β Peg	M2.5II–III
β And	M0III
β Gem	K0III
α Hya	K3II–III

Note: This is taken from Cohen and co-workers (1992–1996).

4.6.1 Hot interstellar medium

Most space in the galaxy is filled with hot ionized matter at very low density ($n \sim 5 \times 10^{-3}$ cm^{-3}) and is heated to very high temperatures ($\sim 5 \times 10^5$ K) by the interstellar radiation field (ISRF). This material causes very little extinction and is known as the *hot interstellar medium*. Ionizing radiation such as far-UV and X-rays pass through it with little extinction.

Within the hot medium, zones of higher density are embedded. Their temperatures and physico-chemical states depend mainly on their densities.

4.6.2 Warm interstellar medium

The lowest density embedded regions are the "warm" clouds with n around 3×10^{-1} cm^{-3} and temperatures of order 10,000 K. They remain transparent enough for their interiors to be heated by the energetic photons of the hot interstellar medium.

4.6.3 Cool interstellar medium

When its density reaches $n = 10^1$–10^2 cm^{-3} the center of a cloud may contain unionized atomic H, though molecules will still have short lifetimes against dissociation. A characteristic temperature is 80 K. These clouds are called *diffuse*. Dust may survive, but not the coolest ices. The outer, unshielded, part of the cloud consists of warm interstellar medium. When a strong UV source is nearby, the diffuse medium is transformed into a *photodissociation region*.

Examples of the relatively refractory substances that survive easily are silicates, with strong spectroscopic signatures at 9.7 and 18.8 μm. There are weaker features at 3.0 and 3.4 μm which may be due to H_2O or CH-containing compounds.

4.6.4 Cold interstellar medium

At still higher densities, $n > 10^2$ cm^{-3}, the cloud cores are shielded from almost all high-energy photons and molecules are abundant. The temperature of the cores is typically 15 K. We are now dealing with *dense clouds* in which ices can form and have a continued existence. The dense core of a cloud is surrounded by a shell of cool interstellar medium which, in turn, is surrounded by a zone of warm interstellar medium.

The spectra of objects seen through molecular clouds include the "refractory" features of the cool clouds described in section 4.6.3 as well as clearly defined water absorption at 3.05 μm and other bands arising from CO, CO_2 and CH_3OH ices (also see section 4.10.2).

4.7 Photodissociation regions

Photodissociation regions (PDRs) exist in the interface zones between molecular clouds and ionized or HII regions. Those with densities $\geq 10^3$ cm^{-3} and $100 < T_{\text{gas}} < 1500$ K are seen to radiate in the forbidden fine-structure lines of [OI] 63 and 146 μm, [CII] 158 μm and [SiII] 35 μm. These lines arise from magnetic dipole radiation. They are dominant because H and He have no suitable low-energy levels which can cool the interstellar gas under these conditions. This process is called *fine-structure cooling*.

A recent review of dense photodissociation regions was given by Hollenbach and Tielens (1997).

4.7.1 [CII] (C^+) in photodissociation regions

The [CII] 158 μm line is one of the strongest and most ubiquitous emission lines of the Milky Way galaxy, where it is the main coolant of the photodissociation regions and is very strongly correlated with dust emission. The heating of the gas is provided by photoelectrons emitted from dust grains under UV bombardment, and their energy is ultimately removed by [CII] emission.

4.7.2 H_2 in photodissociation regions

Cold clouds with $T < 100$ K do not radiate in the rotation-vibration bands of H_2. However, two electronic states above the ground state can be excited by photons of less than 13.6 eV energy (the cut-off energy of the interstellar radiation field) and these are C $^1\Pi_u$ and B $^1\Sigma_u^+$. About 10% of the electronic decays from these states end up in the vibrational continuum states of the ground electronic state, i.e., the molecule comes apart or is *dissociated*. The remainder end up as bound vibrational states which eventually decay (fluoresce) as rotation-vibration transitions.

Low-density gas ($n \leq 10^4$ cm^{-3}) exposed to an ultraviolet field, although remaining around $T \sim 100$ K, is characterized by having the intensity ratio of 2.12 μm 1–0 S(1) to 2.25 μm 2–1 S(1) about 2, which would imply an excitation temperature of $\sim$6000 K (see Table 4.3). This has no relation to the kinetic temperature.

Table 4.3. *H_2 fluorescent emission line intensities*

μm	$100I/I_{tot}$	Line	μm	$100I/I_{tot}$	Line
1.16	0.67	2–0 S(1)	2.12	1.59	1–0 S(1)
1.23	0.85	3–1 S(1)	2.22	0.76	1–0 S(0)
1.31	0.79	4–2 S(1)	2.25	0.89	2–1 S(1)
1.31	0.79	3–1 Q(1)	2.41	1.36	1–0 Q(1)
1.40	0.80	4–2 Q(1)	2.41	0.84	1–0 Q(2)
1.49	0.71	5–3 Q(1)	2.42	1.12	1–0 Q(3)
1.51	0.68	4–2 O(3)	2.55	0.86	2–1 Q(1)
1.83	0.68	1–0 S(5)	2.57	0.65	2–1 Q(3)
1.96	1.26	1–0 S(3)	2.80	1.15	1–0 O(3)
2.03	0.97	1–0 S(2)	2.97	0.74	2–1 O(3)

Notes: The title refers to the fractional intensities of the 20 strongest lines of the H_2 emission spectrum in the 1–4 μm band produced in a photodissociated region with $n_T = 10^3$ cm^{-3} and $\chi = 10^2$, relative to the total intensity $I_{tot} = 4.90 \times 10^{-5}$ erg s^{-1} cm^{-2} s^{-1} (taken from Sternberg, 1990). χ is the factor by which the ultraviolet intensity exceeds the average value for the galaxy.

Table 4.4. H_2 *thermal emission line intensities*

μm	$100I/I_{\text{tot}}$	Line	μm	$100I/I_{\text{tot}}$	Line
1.73	0.16	1–0 S(7)	2.44	0.34	1–0 Q(4)
1.83	0.40	1–0 S(5)	2.45	0.62	1–0 Q(5)
1.89	0.26	1–0 S(4)	2.47	0.12	1–0 Q(6)
1.96	1.27	1–0 S(3)	2.50	0.18	1–0 Q(7)
2.03	0.61	1–0 S(2)	2.63	0.59	1–0 O(2)
2.12	2.17	1–0 S(1)	2.80	2.07	1–0 O(3)
2.22	0.59	1–0 S(0)	3.00	0.50	1–0 O(4)
2.41	2.44	1–0 Q(1)	3.23	0.86	1–0 O(5)
2.41	0.65	1–0 Q(2)	3.50	0.14	1–0 O(6)
2.42	1.53	1–0 Q(3)	3.81	0.17	1–0 O(7)

Notes: The title refers to the fractional intensities of the 20 strongest lines of the H_2 emission spectrum in the 1–4 μm band produced in a photodissociated region with $n_T = 10^6 \, \text{cm}^{-3}$ and $\chi = 10^2$, relative to the total intensity $I_{\text{tot}} = 5.38 \times 10^{-5} \, \text{erg} \, \text{s}^{-1} \, \text{cm}^{-2} \, \text{s}^{-1}$ (taken from Sternberg, 1990). χ is the factor by which the ultraviolet intensity exceeds the average value for the galaxy.

In warm high-density regions ($n \geq 10^4 \, \text{cm}^{-3}$), excitation by UV becomes insignificant compared to thermal collisional excitation. This type of region arises on the boundary of a cold, dense, molecular cloud where the molecules excited by the ultraviolet radiation field are de-excited by collisions. The fine-structure cooling mechanism is also quenched by collisions, and the temperature can reach 10^3 K. The energies are such that the lowest excited level is populated much more than higher levels, and the temperature that can be derived from the ratio of the 1–0 S(1) and 2–1 S(1) lines is close to the kinetic temperature of the gas (Table 4.4). Further diagnostics, such as the ratio of the 2–1 S(1) to the 6–4 Q(1) lines ($\lambda = 1.6015 \, \mu$m), the ortho: para ratio and rotational temperatures are discussed by Draine and Bertoldi (1996).

4.7.2.1 Pure rotational lines of H_2

Pure rotational lines can also be generated in photodissociation regions and shocks. If excited collisionally they do not require such a high-temperature medium as the vibrational lines and can be emitted at 200 K, for example. Their intensities were predicted by Burton, Hollenbach and Tielens (1992) for photodissociation regions of various

Table 4.5. *Wavelengths of pure rotational H_2 lines*

Line	$\lambda(\mu m)$
0–0 S(0)	28.221
0–0 S(1)	17.035
0–0 S(2)	12.279
0–0 S(3)	9.665
0–0 S(4)	8.026
0–0 S(5)	6.908

densities and UV field energy densities. Table 4.5 gives the wavelengths of the first few lines.

A number of these lines have been observed in or near young stellar objects by Wesselius et al. (1996) and Timmermann et al. (1996), using the ISO satellite, several of them for the first time. The rotational population distributions are consistent with thermal excitation by gas at 500–800 K.

4.8 HII regions

An HII region can be thought of as a body of gas, predominantly H, which is kept in an ionized state through UV radiation from hot stars either nearby or embedded within it. Photons with wavelengths shorter than 911.7 Å have enough energy to cause ionization of H atoms and are known as Lyman continuum photons. An ionized atom eventually captures an electron and emits one or more photons as it tries to return to its ground state. Under commonly encountered conditions, the HII region is dense enough so that no ionizing photons escape from it. Their energy is degraded into more numerous lower-energy photons. The assumption that a nebula is optically thick in this way enables its emission to be treated theoretically with results that are found to be quite realistic. This assumption is usually referred to as *Menzel's case B* for historical reasons. A good general reference for HII regions is Osterbrock (1989).

The infrared spectrum of HII regions can, as in the visible, be divided into two parts, the continuum and the line (recombination) spectra. In the infrared, the two-photon continuum, resulting from the decay of the $2\,^2S$ state of hydrogen, is not significant compared to the free-free and free-bound continua.

4.8.1 Free-free and free-bound continua

The acceleration and deceleration of a free electron in the electric field of another charged particle gives rise to radiation, often referred to as *Bremsstrahlung*. If the interaction leaves the electron free, the radiation is called free-free; if the electron becomes bound it is called free-bound.

In an HII region, the line emission is accompanied by continuous radiation from these sources, and both depend on $n_e^2 V$ and temperature, where n_e is the electron density and V is the volume of the region. The ratio of line strength to continuum strength is only a function of temperature.

Free-bound and free-free radiation is the cause of circumstellar emission from Be stars, which have equatorial disk-shaped gaseous stellar envelopes that may become optically thick at $\lambda \geq 2.2\,\mu\mathrm{m}$ (Dachs, Engels and Kiehling, 1988).

The calculation of models for the continua involves approximating the "Gaunt factors," and this has been done by various authors for several different sets of conditions, yielding agreement within about 1%. For example, Ferland (1980) has calculated the HI and HeII continuous emission coeficients for temperatures from $500\,\mathrm{K}$ to $2 \times 10^6\,\mathrm{K}$. The HI emission coefficients are also averaged over commonly used filter bands so that the colors of the continuous emission component can readily be found (Table 4.6).

Table 4.6. *JHKL colors of HI continuous emission at various temperatures*

Temperature (K)	J–H	H–K	K–L
1000	−1.11	2.28	0.30
2500	−0.24	1.16	0.64
5000	0.12	0.82	0.79
10,000	0.32	0.68	0.86
20,000	0.44	0.61	0.90
50,000	0.50	0.58	0.94
500,000	0.53	0.55	0.93

Note: This is calculated from Ferland (1980), Table III, which gives the emission coefficients averaged over the infrared filters in erg $\mathrm{s}^{-1}\,\mathrm{cm}^3\,\mathrm{Hz}^{-1}$.

4.8.2 The recombination spectrum

Collisional processes play no significant part in the depopulation of excited levels in normal HII regions. Only in the ultra-dense conditions which occur in the broad-line region of the nuclei of active galaxies do they become important.

The ratios of the hydrogen Balmer line strengths to Hβ have been calculated and tabulated by Brocklehurst (1971) for several temperatures and electron densities likely to be encountered in HII regions, together with Paschen-to-Balmer intensity ratios and the corresponding ratios for the He$^+$ lines. The dependence of these ratios on the electron density is so much weaker than on the temperature that it can normally be ignored. This work was extended to Brackett-to-Balmer line ratios by Giles (1977). Further extensions to cover a greater range of physical conditions are given by Hummer and Storey (1987).

The ratio of line to continuum flux from an HII region is easily obtained. The emission in the Brγ line is obtained from Table 4.7 as

Table 4.7. *HI recombination line ratios (case B)*

T	5000 K		10,000 K			20,000 K	
n_e (cm^{-3})	10^2	10^4	10^2	10^4	10^6	10^2	10^4
Balmer-line intensities relative to Hβ							
$I_{H\alpha}/I_{H\beta}$	3.04	3.00	2.86	2.85	2.81	2.75	2.74
$I_{H\gamma}/I_{H\beta}$	0.458	0.460	0.468	0.469	0.471	0.475	0.476
$I_{H\delta}/I_{H\beta}$	0.251	0.253	0.259	0.260	0.262	0.264	0.264
$I_{H\epsilon}/I_{H\beta}$	0.154	0.155	0.159	0.159	0.163	0.163	0.163
$I_{H8}/I_{H\beta}$	0.102	0.102	0.105	0.105	0.110	0.107	0.107
Brackett-line intensities relative to corresponding Balmer lines							
$I_{Br\alpha}/I_{H\gamma}$	0.227	0.215	0.171	0.166	0.154	0.132	0.127
$I_{Br\beta}/I_{H\delta}$	0.222	0.214	0.175	0.172	0.163	0.141	0.140
$I_{Br\gamma}/I_{H\epsilon}$	0.214	0.209	0.175	0.173	0.163	0.144	0.143
$I_{Br\delta}/I_{H8}$	0.209	0.206	0.174	0.172	0.160	0.146	0.145

Notes: For more ratios, the reader may consult Brocklehurst (1971) and Giles (1977) or Table 4.4 of Osterbrock (1989). Also, $I_{H\beta}$ for a 10^4 K HII region is given by $4\pi I_{H\beta} \sim 1.24 \times 10^{-25}$ erg cm^3 sec^{-1} (Osterbrock, 1989; Table 4.4).

follows:

$$4\pi I_{\text{Br}\gamma} = \frac{I_{\text{Br}\gamma}}{I_{\text{He}}} \frac{I_{\text{He}}}{I_{\text{H}\beta}} I_{\text{H}\beta} n_e^2 V = 3.4 \times 10^{-27} n_e^2 V \text{ erg cm}^3 \text{ sec}^{-1}$$

for $T = 10,000$ and $n_e = 10^4 \text{ cm}^3$.

Similarly, the continuum emission coefficient j_ν at the wavelength of Brγ can be read from Ferland (1980) and is given by $4\pi j_\nu \sim 7.8 \times 10^{-40}$ $n_e^2 V \text{ erg cm}^3 \text{ sec}^{-1} \text{ Hz}^{-1}$ at 10,000 K.

As already stated, the overall emission of an HII region is determined by the product $n_e^2 V$, where n_e (cm^{-3}) is the electron density and V (cm^3) is the volume of the HII region. This quantity is related to the number of ionizing Lyman continuum photons, for case B and $T = 10,000$ K, in a simple hydrogen nebula of volume V, by the expression

$$N(\text{Lyc}) \sim 2.6 \times 10^{-13} \, n_e^2 V \text{ photons s}^{-1} (\text{Osterbrock, 1989; Table 2.1.}).$$

The Brackett α luminosity is given by

$$L(\text{Br}\alpha) = \frac{L_{\text{Br}\alpha}}{L_{\text{H}\gamma}} \frac{L_{\text{H}\gamma}}{L_{\text{H}\beta}} L_{\text{H}\beta} = 9.46 \times 10^{-34} \left(\frac{T}{10^4}\right)^{-1.2} n_e^2 V \text{ W},$$

using the results of Brocklehurst (1971) and Giles (1977) and expressing the temperature dependence as a power law.

From the observed of intensity of an infrared line we can thus determine the strength of the Lyman continuum.

Even in the infrared, however, the observed line strength may be affected by interstellar absorption. The radio continuum is not absorbed, so we can use the results of Mezger and Henderson (1967; eqn. A.13) to obtain a better approximate value for n_e, the average electron density in an HII region, from its radio flux (F_ν) at frequency ν.

$$\left(\frac{n_e}{\text{cm}^{-3}}\right) = 6.4 \times 10^2 \left(\frac{T}{10^4}\right)^{0.175} \left(\frac{\nu}{\text{GHz}}\right)^{0.05} \left(\frac{F_\nu}{\text{Jy}}\right)^{0.5} D^{-0.5} \theta^{-1.5}.$$

(D is its distance in kpc and θ is its angular size in arcmin.)

The expected Brα line intensity will then be

$$I(\text{Br}\alpha) = 2.71 \times 10^{-14} \left(\frac{T}{10^4}\right)^{-0.85} \left(\frac{\nu}{5\,\text{GHz}}\right)^{0.1} F_\nu(\text{Jy}) \text{ W m}^{-2}$$

(see Moorwood and Salinari, 1983). This approach can be very useful in estimating the interstellar extinction in front of an HII region, by comparing the 5 GHz radio flux with the observed numbers of photons in the Brα or any other measured line.

A tabulation by Panagia (1973) gives the number of Lyman contin-uum photons emitted by early-type stars according to spectral type and luminosity class. Their absolute bolometric luminosities are also given. These quantities depend very steeply on spectral type. For example, a B3 zero-age main sequence star will emit $10^{43.69}$ Lyc photons sec^{-1} ($T_{\text{eff}} = 17,900 \, \text{K}$) and have a luminosity $\log L/L_{\odot} = 3.02$, while a ZAMS O4 star will emit $10^{49.93}$ Lyc photons sec^{-1} ($T_{\text{eff}} = 50,000 \, \text{K}$) and have $\log L/L_{\odot} = 6.11$.

The spectral type of a ZAMS star surrounded by dust in a compact HII region can be estimated roughly from its 5 GHz continuum flux and bolometric magnitude. The bolometric flux, mainly emerging in the infrared, should be equal to that of the star, while the 5 GHz continuum is an indicator of the number of Lyman continuum photons not absorbed by dust, which may be from 10% to 50% of the total. Even though the total number of Lyc photons may be uncertain by up to a factor of 5, the steepness of the luminosity dependence constrains the spectral type quite closely (Wood and Churchwell, 1989). This argument can be extended to estimate the upper temperature cutoff of the distribution of hot stars in a starburst region. The relevance of this to the initial mass function (IMF) will depend on whether the starburst is ongoing or how old it is. The ratio of the bolometric luminosity of an HII region to, for example, its Lyα luminosity fixes the predominant spectral type of the exciting stars. Making use of the stellar mass–luminosity relation then gives their mass.

4.8.3 Fine-structure lines

In addition to the hydrogen lines, fine-structure forbidden lines are present in spectra from HII regions, planetary nebulae and many ac-tive galaxies (see Fig. 4.1 and Tables 4.8 and 4.10).

Simpson (1975) gives predictions of line strengths and how they may be used as diagnostics of the physical state of the gas, giving as examples the Orion nebula and some planetary nebulae.

Atoms in the upper level of a possible transition can be de-excited either by the emission of an observable photon or by collisions. There is a certain *critical density* $n_e(\text{crit})$ of electrons for each transition above which collision de-excitation predominates over radiation. For $n_e \gg n_e$ (crit), where only collisional processes need be considered, thermal equi-librium sets in and the line flux is proportional to $n_i dV$, the total number

Table 4.8. *Some emission lines observed in the 1–2.5 μm region*

λ_{air}	Species	Transition	LkHα	H	η	C	S
1.0049	HI	Paδ	*		*		*
1.0129	HeII	possibly + Ar XIII					*
1.0174	FeII	$b^4G_{9/2}$–$z^4D^o_{7/2}$	*		*		
1.0310	He	+ [SII] and/or OI			*		*
1.0406	[NI]	$3F$			*		
1.0501	FeII	$b^4G_{9/2}$–$z^4F^o_{7/2}$ +	*		*		
1.0747	[FeXIII]						*
1.0798	[FeXIII]						*
1.0830	HeI	$2p^3P^o$–$2s^3S$	*		*		*
1.0863	FeII	$b^4G_{7/2}$–$z^4F^o_{5/2}$	*		*		
1.0938	HI	Paγ	*		*		*
1.1126	FeII	$b^4G_{5/2}$–$z^4F^o_{3/2}$	*		*		
1.1287	OI	$3d^3D^o$–$3p^3P$	*				
1.2528	[SiX] +	blend with HeI					*
1.2567	[FeII]	$a^4D_{7/2}$–$a^6D_{9/2}$	*		*		*
1.279	[FeII]	$a^4D_{3/2}$–$a^6D_{3/2}$					*
1.2820	HI	Paβ	*		*		*
1.3164	OI		*				
1.3209	[FeII]	$a^4D_{7/2}$–$a^6D_{7/2}$					*
1.328	[FeII]	$a^4D_{5/2}$–$a^6D_{3/2}$					*
1.5335	[FeII]	$a^4D_{5/2}$–$a^4F_{9/2}$	*	*	*		*
1.5530	FeII	z^4P–c^2P	*				
1.5995	[FeII]	$a^4D_{3/2}$–$a^4F_{7/2}$		*	*		
1.6436	[FeII]	$a^4D_{7/2}$–$a^4F_{9/2}$	*	*	*		*
1.6638	[FeII]	$a^4D_{1/2}$–$a^4F_{5/2}$		*	*		*
1.6769	[FeII]	$a^4D_{5/2}$–$a^4F_{7/2}$	*	*	*		*
1.6787	FeII	c^4F_7–z^4F_9	*		*		
1.6871	FeII	blend	*		*		
1.6996	HeI	$4d^3D$–$3p^3P^o$		*	*		
1.7338	FeII	c^4P_5–z^4D_7	*				
1.7362	FeII	z^4D–c^4F	*				
1.7414	FeII	c^4F_7–z^4F_7	*		*		
1.875	HI	Paα					*
1.957	H$_2$	(1–0) S(3)					*
1.9615	[SiVI]	$^2P_{1/2}$–$^2P_{3/2}$					*
1.9742	FeII	c^4F_5–z^4F_5			*		
2.0337	H$_2$	(1–0) S(2)					*
2.0580	HeI	$2p^1P$–$2s^1S$		*	*		
2.0886	FeII	c^4F_3–z^4F_3			*		
2.1115	HeI	$4s^3S$–$3p^3P^o$		*	*		
2.121	H$_2$	(1–0) S(1)		*			*
2.1369	MgII	$5p^2P^o_{3/2}$–$5s^2S_{1/2}$		*			
2.1432	MgII	$5p^2P^o_{1/2}$–$5s^2S_{1/2}$		*			
2.1445	[CrII]	a^2H–a^2I?	*				

Table 4.8. *(cont.)*

λ_{air}	Species	Transition	LkHα	H	η	C	S
2.1451	[FeIII]	$^3G_3$–$^3H_4$				*	
2.1654	HI	Brγ	*		*	*	*
2.205	[FeXII]	$^2D_{5/2}$–$^2D_{3/2}$				*	
2.2178	[FeIII]	$^3G_5$–$^3H_6$				*	
2.223	H_2	(1–0) S(0)		*			*
2.2420	[FeIII]	$^3G_4$–$^3H_4$				*	
2.247	H_2	(2–1) S(1)		*			*
2.321	[CaVIII]						*
2.3479	[FeIII]	$^3G_5$–$^3H_5$		*		*	
2.4827	[SiVII]	$^3P_1$–$^3P_2$				*	*

Notes: Not all spectra cover the whole range. In particular, atmospheric absorption bands are omitted. Wavelengths and transition designations are taken from the papers referred to. Many HI and H_2 lines are omitted. LkHα is the protostellar source LkHα101, from Rudy et al. (1991), Simon and Cassar (1984) and Hamann and Persson (1989). H is the young planetary nebula Hubble 12, from Luhman and Rieke (1996); most HI and H_2 lines are excluded from the list. η is the peculiar old nova η Car, from Hamann et al. (1994); only lines with flux $> 10^{-10}\,\text{erg}\,\text{cm}^{-2}\,\text{sec}^{-1}$ are included; some H lines are excluded. C is the Galactic Centre mini-cavity, from Lutz, Krabbe and Genzel (1993). S includes lines from the Seyfert galaxies NGC 1068 (Oliva and Moorwood, 1990, and Marconi et al., 1996) and NGC 4151 (Thompson, 1995).

of ions in the emitting volume dV. When radiative de-excitation predominates, the line flux is proportional to $n_i n_e dV$.

Ions with p^1 and p^5 outer electronic configurations have doublet fine-structure splitting and so have only one fine-structure transition in their ground states. Examples of this are [SIV] at $10.5\,\mu$m and [NeII] at $12.8\,\mu$m.

Those with p^2 and p^4 configurations will have triplet ground terms giving rise to pairs of infrared lines. An example of this is [OIII] at 52 and $88\,\mu$m. For certain ranges of n_e the critical densities of one line may be proportional to n_e and the other to $n_i n_e$, so that n_i may be determined.

Table 4.9, taken from Barlow (1989), gives some fine-structure line-pairs which cover a large range in critical densities. More extensive atomic data are available from Stacey (1989) and Spinoglio and Malkan (1992).

Electron temperatures may be determined from infrared fine-structure lines and visible forbidden lines from the same ion. This requires detailed

Table 4.9. *Critical densities for some infrared fine-structure lines*

Ion	λ (μm)	n_e(crit)(m^{-3})
NII	121.7	2.5×10^7
	203.9	2.5×10^8
OIII	51.82	6×10^8
	88.36	5×10^9
SIII	33.47	4×10^9
	18.71	4×10^{10}
NeIII	36.02	3×10^{10}
	15.56	4×10^{11}
ArIII	21.83	4×10^{10}
	8.99	4×10^{11}
NeV	24.28	1×10^{11}
	14.33	4×10^{11}

Note: This is taken from Barlow (1989).

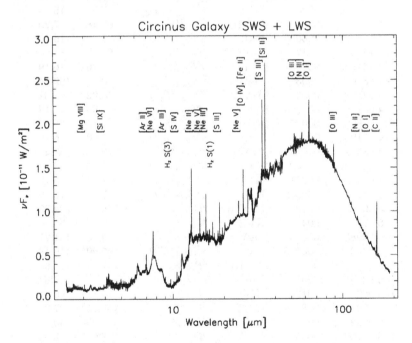

Fig. 4.1. ISO spectrum of the Circinus galaxy. This complex object emits fine-structure lines of high excitation characteristic of Seyfert galaxies as well as low excitation lines which show the simultaneous presence of a starburst. Also visible are the PAH features, H_2 molecular emission lines and a far-infrared continuum peak; from Moorwood (1997).

Table 4.10. *Atomic and ionic fine-structure lines in the range 2–205 µm, covered by the ISO satellite*

λ	Err_λ	Species	Transition	EP	IP
1.9634	.0001	[SiVI]	$^2P_{1/2}-^2P_{3/2}$	166.77	205.05
2.4833	.0001	[SiVII]	$^3P_1-^3P_2$	205.05	246.52
2.58423	.00023	[SiIX]	$^3P_2-^3P_1$	303.17	351.10
2.90519	.00037	[AlV]	$^2P_{1/2}-^2P_{3/2}$	120.00	153.83
2.9846	.0001	[CoII]	$a^5F_5-a^3F_4$	7.86	17.08
3.02795	.00025	[MgVIII]	$^2P_{3/2}-^2P_{1/2}$	224.95	265.96
3.1200	.0001	[NiI]	$a^1D_2-a^3D_3$		7.64
3.19053	.00015	[KVII]	$^2P_{3/2}-^2P_{1/2}$	99.40	117.56
3.20671	.00015	[CaIV]	$^2P_{1/2}-^2P_{3/2}$	50.91	67.27
3.660	.001	[AlVI]	$^3P_1-^3P_2$	153.83	190.48
3.690	.014	[AlVIII]	$^3P_2-^3P_1$	241.44	284.60
3.9357	.0004	[SiIX]	$^3P_1-^3P_0$	303.17	351.10
3.9524	.0001	[NiI]	$a^1D_2-a^3D_2$		7.64
4.076319	.000004	[FeII]	$a^4F_{5/2}-a^6D_{7/2}$	7.90	16.19
4.081906	.000004	[FeII]	$a^4F_{3/2}-a^6D_{5/2}$	7.90	16.19
4.087	.005	[CaVII]	$^3P_2-^3P_1$	108.78	127.20
4.114990	.000005	[FeII]	$a^4F_{7/2}-a^6D_{9/2}$	7.90	16.19
4.1585	.0017	[CaV]	$^3P_1-^3P_2$	67.27	84.50
4.434839	.000007	[FeII]	$a^4F_{3/2}-a^6D_{3/2}$	7.90	16.19
4.48668	.0003	[MgIV]	$^2P_{1/2}-^2P_{3/2}$	80.14	109.24
4.52952	.00031	[ArVI]	$^2P_{3/2}-^2P_{1/2}$	75.02	91.01
4.607664	.000005	[FeII]	$a^4F_{5/2}-a^6D_{5/2}$	7.90	16.19
4.61802	.00047	[KIII]	$^2P_{1/2}-^2P_{3/2}$	31.63	45.80
4.671945	.000005	[FeII]	$a^4F_{3/2}-a^6D_{1/2}$	7.90	16.19
4.68472	.00033	[NaVII]	$^2P_{3/2}-^2P_{1/2}$	172.15	208.44
4.889137	.000005	[FeII]	$a^4F_{7/2}-a^6D_{7/2}$	7.90	16.19
5.06235	.00001	[FeII]	$a^4F_{5/2}-a^6D_{3/2}$	7.90	16.19
5.340169	.000009	[FeII]	$a^4F_{9/2}-a^6D_{9/2}$	7.90	16.19
5.5032	.0012	[MgVII]	$^3P_2-^3P_1$	186.51	224.95
5.575	.004	[KVI]	$^3P_2-^3P_1$	82.66	99.40
5.60985	.00063	[MgV]	$^3P_1-^3P_2$	109.24	141.27
5.673905	.000008	[FeII]	$a^4F_{7/2}-a^6D_{5/2}$	7.90	16.19
5.85	.04	[AlVIII]	$^3P_1-^3P_0$	241.44	284.60
5.8933	.0001	[NiI]	$a^1D_2-a^3D_1$		7.64
5.9820	.0012	[KIV]	$^3P_1-^3P_2$	45.80	60.91
6.154	.008	[CaVII]	$^3P_1-^3P_0$	108.78	127.20
6.4922	.0011	[SiVII]	$^3P_0-^3P_1$	205.05	246.52
6.6360	.0003	[NiII]	$^2D_{3/2}-^2D_{5/2}$	7.64	18.17
6.704	.009	[ClV]	$^2P_{3/2}-^2P_{1/2}$	53.46	67.82
6.721283	.000011	[FeII]	$a^4F_{9/2}-a^6D_{7/2}$	7.90	16.19
6.985274	.000003	[ArII]	$^2P_{1/2}-^2P_{3/2}$	15.76	27.63
7.3177	.0012	[NaIII]	$^2P_{1/2}-^2P_{3/2}$	47.29	71.62
7.5067	.0002	[NiI]	$a^3F_3-a^3F_4$		7.64

Table 4.10. (cont.)

λ	Err$_\lambda$	Species	Transition	EP	IP
7.6524	.0012	[NeVI]	$^2P_{3/2}-^2P_{1/2}$	126.21	157.93
7.8145	.0012	[FeVII]	$^3F_4-^3F_3$	99.10	124.98
7.9016	.0002	[ArV]	$^3P_2-^3P_1$	59.81	75.02
8.61059	.00088	[NaVI]	$^3P_2-^3P_1$	138.39	172.15
8.8299	.0002	[KVI]	$^3P_1-^3P_0$	82.66	99.40
8.99138	.00012	[ArIII]	$^3P_1-^3P_2$	27.63	40.74
9.0090	.0004	[MgVII]	$^3P_1-^3P_0$	186.51	224.95
9.041	.001	[NaIV]	$^3P_1-^3P_2$	71.62	98.91
9.116	.008	[AlVI]	$^3P_0-^3P_1$	153.83	190.48
9.5267	.0011	[FeVII]	$^3F_3-^3F_2$	99.10	124.98
10.5105	.0001	[SIV]	$^2P_{3/2}-^2P_{1/2}$	34.79	47.22
10.521	.001	[CoII]	$a^3F_3-a^3F_4$	7.86	17.08
10.6822	.0008	[NiII]	$^4F_{7/2}-^4F_{9/2}$	7.64	18.17
11.3075	.0002	[NiI]	$a^3F_2-a^3F_3$		7.64
11.333347	.000015	[CII]	$^2P_{1/2}-^2P_{3/2}$		12.97
11.482	.019	[CaV]	$^3P_0-^3P_1$	67.27	84.50
11.741	.007	[ClIV]	$^3P_2-^3P_1$	39.61	53.46
11.89	.01	[CoIII]	$a^4F_{7/2}-a^4F_{9/2}$	17.08	33.50
12.0010	.0003	[NiI]	$a^3D_1-a^3D_2$		7.64
12.255	.002	[CoI]	$a^4F_{7/2}-a^4F_{9/2}$		7.86
12.7288	.0012	[NiII]	$^4F_{5/2}-^4F_{7/2}$	7.64	18.17
12.81355	.00002	[NeII]	$^2P_{1/2}-^2P_{3/2}$	21.56	40.96
13.1022	.0002	[ArV]	$^3P_1-^3P_0$	59.81	75.02
13.432	.009	[FV]	$^2P_{3/2}-^2P_{1/2}$	87.14	114.24
13.5213	.0002	[MgV]	$^3P_0-^3P_1$	109.24	141.27
14.3217	.0002	[NeV]	$^3P_2-^3P_1$	97.12	126.21
14.3678	.0008	[ClIII]	$^3P_1-^3P_2$	12.97	23.81
14.3964	.0021	[NaVI]	$^3P_1-^3P_0$	138.39	172.15
14.740	.002	[CoII]	$a^5F_4-a^5F_5$	7.86	17.08
14.8142	.0003	[NiI]	$a^3D_2-a^3D_3$		7.64
15.155	.002	[CoI]	$b^4F_{7/2}-b^4F_{9/2}$		7.86
15.39	.03	[KIV]	$^3P_0-^3P_1$	45.80	60.91
15.460	.002	[CoII]	$a^3F_2-a^3F_3$	7.86	17.08
15.5551	.0003	[NeIII]	$^3P_1-^3P_2$	40.96	63.45
16.39	.03	[CoIII]	$a^4F_{5/2}-a^4F_{7/2}$	17.08	33.50
16.925	.003	[CoI]	$a^4F_{5/2}-a^4F_{7/2}$		7.86
17.885	.005	[PIII]	$^2P_{3/2}-^2P_{1/2}$	19.77	30.20
17.93595	.00009	[FeII]	$a^4F_{7/2}-a^4F_{9/2}$	7.90	16.19
18.2405	.0020	[NiII]	$^4F_{3/2}-^4F_{5/2}$	7.64	18.17
18.264	.003	[CoI]	$b^4F_{5/2}-b^4F_{7/2}$		7.86
18.7130	.0002	[SIII]	$^3P_2-^3P_1$	23.34	34.79
18.804	.004	[CoII]	$a^5F_3-a^5F_4$	7.86	17.08
20.354	.021	[ClIV]	$^3P_1-^3P_0$	39.61	53.46

Table 4.10. *(cont.)*

λ	Err_λ	Species	Transition	EP	IP
21.29	.06	[NaIV]	$^3P_0-^3P_1$	71.62	98.91
21.829	.002	[ArIII]	$^3P_0-^3P_1$	27.63	40.74
22.925	.002	[FeIII]	$^5D_3-^5D_4$	16.19	30.65
24.04233	.00006	[FeI]	$a^5D_3-a^5D_4$		7.90
24.07	.06	[CoIII]	$a^4F_{3/2}-a^4F_{5/2}$	17.08	33.50
24.3175	.0003	[NeV]	$^3P_1-^3P_0$	97.12	126.21
24.51925	.00016	[FeII]	$a^4F_{5/2}-a^4F_{7/2}$	7.90	16.19
24.7475	.0015	[FI]	$^2P_{1/2}-^2P_{3/2}$		17.42
24.845	.006	[CoI]	$a^4F_{3/2}-a^4F_{5/2}$		7.86
25.2490	.0003	[SI]	$^3P_1-^3P_2$		10.36
25.681	.007	[CoII]	$a^5F_2-a^5F_3$	7.86	17.08
25.83	.04	[FIV]	$^3P_2-^3P_1$	62.71	87.14
25.8903	.0003	[OIV]	$^2P_{3/2}-^2P_{1/2}$	54.93	77.41
25.930	.007	[CoI]	$b^4F_{3/2}-b^4F_{5/2}$		7.86
25.98829	.00018	[FeII]	$a^6D_{7/2}-a^6D_{9/2}$	7.90	16.19
29.33	.04	[FII]	$^3P_1-^3P_2$	17.42	34.97
32.87	.03	[PII]	$^3P_2-^3P_1$	10.48	19.77
33.0384	.0066	[FeIII]	$^5D_2-^5D_3$	16.19	30.65
33.281	.008	[ClII]	$^3P_0-^3P_1$	12.97	23.81
33.4810	.0003	[SIII]	$^3P_1-^3P_0$	23.34	34.79
34.71330	.00012	[FeI]	$a^5D_2-a^5D_3$		7.90
34.8152	.0005	[SiII]	$^2P_{3/2}-^2P_{1/2}$	8.15	16.35
35.34865	.00026	[FeII]	$a^6D_{5/2}-a^6D_{7/2}$	7.90	16.19
35.77740	.00037	[FeII]	$a^4F_{3/2}-a^4F_{5/2}$	7.90	16.19
36.0135	.0003	[NeIII]	$^3P_0-^3P_1$	40.96	63.45
39.274	.015	[CoII]	$a^5F_1-a^5F_2$	7.86	17.08
44.07	.21	[FIV]	$^3P_1-^3P_0$	62.71	87.14
51.30044	.00091	[FeII]	$a^6D_{3/2}-a^6D_{5/2}$	7.90	16.19
51.68	.2	[FeIII]	$^5D_1-^5D_2$	16.19	30.65
51.8145	.0005	[OIII]	$^3P_2-^3P_1$	35.12	54.93
54.31093	.00042	[FeI]	$a^5D_1-a^5D_2$		7.90
56.31	.01	[SI]	$^3P_0-^3P_1$		10.36
57.317	.002	[NIII]	$^2P_{3/2}-^2P_{1/2}$	29.60	47.45
60.64	.07	[PII]	$^3P_1-^3P_0$	10.48	19.77
63.183705	.000002	[OI]	$^3P_1-^3P_2$		13.62
67.2	.3	[FII]	$^3P_0-^3P_1$	17.42	34.97
68.473	.004	[SiI]	$^3P_2-^3P_1$		8.15
87.3844	.0027	[FeII]	$a^6D_{1/2}-a^6D_{3/2}$	7.90	16.19
88.3560	.0023	[OIII]	$^3P_1-^3P_0$	35.12	54.93
89.237	.008	[AlI]	$^2P_{3/2}-^2P_{1/2}$		5.99
105.37	.6	[FeIII]	$^5D_0-^5D_1$	16.19	30.65

Table 4.10. *(cont.)*

λ	Err$_\lambda$	Species	Transition	EP	IP
111.1828	.0018	[FeI]	a^5D_0–a^5D_1		7.90
121.89757	.00002	[NII]	$^3P_2$–$^3P_1$	14.53	29.60
129.68173	.00003	[SiI]	$^3P_1$–$^3P_0$		8.15
145.525439	.000007	[OI]	$^3P_0$–$^3P_1$		13.62
157.7409	.0001	[CII]	$^2P_{3/2}$–$^2P_{1/2}$	11.26	24.38
205.17823	.00002	[NII]	$^3P_1$–$^3P_0$	14.53	29.60

Notes: λ is the vacuum wavelength in microns. Err$_\lambda$ is the best information on the uncertainty of this wavelength. EP is the excitation potential in eV to create the relevant species. IP is the ionization potential in eV to get to the next ionization stage. The compiler, Dieter Lutz, expresses special thanks to S. Colgan, M. Haas, S. Lord, G. Nave, and E. Serabyn for providing unpublished data. See also new wavelength information derived from ISO data in Feuchtgruber et al. (1997). This is taken from the MPE home page, with permission. For further lists, see www.mpe.mpg.de/www_ir/ISO/linelists/index.html.

knowledge of collision strengths. The available atomic data were summarized by Barlow (1989). Herter (1989) gives the wavelengths and ionization potentials for many infrared forbidden lines and also discusses how nebular parameters may be determined.

4.8.3.1 [FeII] and [FeIII] lines

[FeII] lines, which occur in a large variety of astrophysically interesting objects (see Table 4.8), may also be useful diagnostics of electron density n_e and temperature T_e (Pradhan and Zhang, 1993). As an example, the [FeII] and [FeIII] lines of the young planetary nebula Hubble 12, which shows fluorescent excitation of H_2, have been discussed by Luhman and Rieke (1996).

Greenhouse et al. (1991) suggest, on observational grounds, that the [FeII] emission of galaxies traces their supernova activity and is independent of their content of HII and photodissociation regions. It originates instead in shocked material. Shock-excited [FeII] lines at 1.26 and 1.64 μm arise from the same level and can be used for determining extinction.

[FeIII] emission at 2.217 μm from the "mini-cavity" at the Galactic Center has been discussed by Lutz, Krabbe and Genzel (1993), who consider its excitation and conclude that it comes from photoionized material with $T_e \sim 7000\,$K.

4.8.4 Numerical modelling of HII regions

Because of the complexity of treating a real HII region, numerical models
find favor. Among these, the best known is "CLOUDY" (Ferland, 1996),
which deals with material around a hot source in an essentially spherical
distribution. The computer code incorporates atomic and molecular
data for vast numbers of lines. Parameters which can be set include the
luminosity and spectral shape of the central source, the density, chemical
composition and fill factor of the clouds. The output is a spectrum
which can be compared with one that has been observed. CLOUDY has
applicability ranging from the intergalactic medium to the broad-line
region of active galaxies.

4.9 Shocks

Shocks arise when material moving faster than the speed of sound encounters a slower-moving medium. The "speed of sound" means here
the speed at which a compressional disturbance is propagated, with gas
pressure as the restoring force. It is typically about 1 km s^{-1} in dense
molecular cloud material (Chernoff and McKee, 1990).

An example occurs in star-forming regions, when a jet-like outflow
from a protostar (see section 5.9) encounters a dense molecular cloud,
giving rise to a Herbig–Haro object. Also, the expansion of a compact
HII region, forming around a young massive hot star, causes a rapid
outflow of material which gives rise to a shocked zone at the boundary
where it encounters the surrounding cooler medium. The expansion of a
supernova remnant into the interstellar medium is another well-known
case.

Shocks are divided into two classes.

C-shocks (continuous) occur when the difference in velocity is relatively low (order of 50 km s^{-1}) and the kinetic energy of the individual
ions is insufficient to dissociate the neutral medium. The cooling of the
shock zone is primarily in the infrared, for example, in H$_2$ emission. The
ambient magnetic field, which is coupled to the ions, may allow C-shock
conditions to occur at higher velocities.

J-shocks (jump) occur at higher velocities, when the kinetic energy
is sufficient to dissociate the H$_2$. The energy of the shock is dissipated
mainly in the UV, but this component may not be detectable since
the shock zone is likely to be inside a dense molecular cloud. Again,
molecular H$_2$ emission may betray its presence. The line ratios observed

are likely to be similar to those observed from fluorescent emission in photodissociation regions, but their widths will be greater because of the velocity difference across and within the shock.

Geballe (1990) summarizes the observational situation concerning H_2 lines.

A series of numerical radiative transfer codes called "MAPPINGS" deals with shocks. MAPPINGS 3 contains the infrared lines and is described by Sutherland et al. (1999).

Fine-structure lines with ionization potentials >50 eV are seen in Seyfert galaxies, but not in starbursts. Lines associated with high ionization potentials are referred to as "coronal," because of their occurrence in the Sun's corona, where they are collisionally excited in shocks. Suitable conditions for their generation also arise just outside the broad-line regions of active galactic nuclei. These regions have been modelled by Ferguson, Korista and Ferland (1997).

4.10 Solid-state features

Molecules are no longer free to rotate in the solid state. Although the observed bands are of vibrational origin, they are broadened and their wavelengths are usually shifted from the values observed in the gaseous state. This is a consequence of their closer packing and of interference by other substances constituting the *matrix*, or agglomeration of grain materials. The shapes of the particles also have an effect.

4.10.1 Silicon compounds

Circumstellar silicate features are seen at 9.7 μm (Si–O stretching) and 18.5 μm (O–Si–O bending) in M stars and some planetary nebulae. They also appear in absorption in the interstellar medium. They appear in emission from circumstellar shells when they are optically thin and in absorption when they are thick.

Silicon carbide (SiC) at 11.2 μm appears usually in emission in C-rich red giants (see Fig. 5.3) and planetary nebulae. However, it does not appear in the general interstellar medium, indicating that most Si is bonded with O.

The presence of the silicate and SiC features is well-correlated with optical evidence for oxygen-richness and carbon-richness, respectively. As discussed in the case of stellar atmospheres, this is because CO forms much more easily and at higher temperatures than other compounds, taking up all the available O and C until nothing is left of one or the other. The remaining, more abundant, one of the two then determines whether the medium is O- or C-rich.

4.10.1.1 Amorphous silicates

A commonly observed feature of the interstellar absorption curve is that at 9.7 μm due to amorphous silicate material. It is accompanied by another band at 18 μm. The strength of the 9.7 μm band is found to be proportional to A_V for interstellar clouds in the solar neighborhood (Whittet, 1987)

$$\frac{A_V}{\tau_{9.7}} = 18.5 \pm 1,$$

but is anomalously high in the direction of the Galactic Center ($A_V/\tau_{9.7} = 9 \pm 1$; Roche and Aitken, 1985).

4.10.1.2 Crystalline silicates

Features seen in the 20–45 μm spectra of dust shells of O-rich objects are believed to arise from crystalline silicates such as olivine and pyroxenes. See Tielens et al. (1988) for a recent review.

4.10.2 Ices

In the dense regions of molecular clouds, where dust grains are sheltered from the interstellar UV radiation field, various types of ices can form on the outer layers of grains. These include water, CO, CO_2, CH_3OH and CH_4 (Fig. 4.2; Table 4.11). The formation of the more complex molecules in some cases is caused by UV irradiation of the simple ones.

Ices are also seen in the spectra of comets. A brief review of interstellar ices has been given by Whittet et al. (1996), with the spectrum of the obscured young stellar object NGC 7538-IRS 9 as an example.

4.10.2.1 CO ice

CO in gaseous form is an important constituent of molecular clouds. When conditions are sufficiently cold (≤ 17 K) it may condense as a frost onto grains. This may cause significant depletion of CO from the gas phase. The shape of the CO absorption feature depends on the nature of the mixture of ices in which it is found, and its width is related to the dipole moment of the predominant host substance. CO frosts may be annealed by excursions to moderate temperature (~ 100 K), causing a decrease in the width of the feature. However, CO ice only survives at

Table 4.11. *Interstellar ices*

Compound	Wavelength	Mode	Reference
CH_4	3.32 (ν_3), 7.66 (ν_4)	str, def.	Boogert et al. (1996)
CO_2	4.27 (ν_3), 15.2 (ν_2)	str, bend	de Graauw et al. (1996)
CO	4.67		
$^{13}CO_2$	4.38	str	
CH_3OH	3.54 (ν_3), 3.47	C–H str	Allamandola et al. (1992)
H_2O	3.0 (ν_1, ν_2), 6.0 (ν_2)	str, bend	
"XCN"	4.62		
?	6.8		

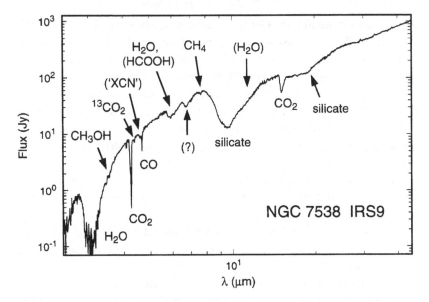

Fig. 4.2. ISO spectrum from 2.4–45 μm of the dust-embedded YSO (young stellar object) NGC 7538-IRS 9, showing absorption bands due to various ices (Whittet et al., 1996).

this temperature as an impurity in H_2O ice. A review has been written by Whittet and Duley (1991).

4.10.2.2 CO_2 ice

CO_2, although very widely observed in solid form, has not been found as a gas in the interstellar medium (van Dishoeck et al., 1996).

As is the case for CO, the shape of the CO_2 feature depends on whether the matrix of ices is polar or non-polar, i.e., whether it is dominated by H_2O or not (de Graauw et al., 1996).

Laboratory simulations of CO_2 ice under varied conditions are given by Ehrenfreund et al. (1996).

4.10.2.3 "XCN"

A feature at 4.62 μm is attributed to $C \equiv N$ bonds in an otherwise unidentified molecule.

4.10.3 The PAH bands

An important family of infrared bands, sometimes referred to as the "unidentified infrared bands" or UIBs, are centered at 3.3, 6.2, 7.7, 8.6 and 11.3 μm. They have been observed in planetary nebulae, HII regions, reflection nebulae around early-type stars, and in some galaxies. In addition, they contribute to the diffuse emission from the disk of the galaxy (Mattila et al., 1996). A favorite suggestion has been that they are due to C–C and C–H stretching and bending vibrations in polycyclic aromatic hydrocarbons (PAHs) (Puget and Léger, 1989). An alternative explanation, that they are caused by hydrogenated amorphous graphitic particles, is due to Duley and Williams (1988). For convenience, we will generally refer to them as the "PAH" bands, though the question of their identity is not wholly settled.

The PAH compounds are platelets of hexagonal carbon rings of various sizes and should perhaps be regarded as large molecules rather than as particles. The simplest member of this group of compounds is coronene, $C_{24}H_{12}$. The bonds on the outskirts of PAH platelets are usually heavily hydrogenated but this may not always be the case under interstellar conditions. A recent paper (Beintema et al., 1996) maps the details of these features in the spectra of evolved C-rich objects (Fig. 4.3) and identifies them with parts of various PAH compounds. The individual bands are superimposed on pronounced pedestals.

The regions which emit PAH spectra are located in the interfaces between hot stars and molecular clouds, i.e., in the photodissociation regions. It is believed that they are excited by the absorption of single energetic photons. They are seen only in emission.

Variations in the relative strengths of the PAH bands with ambient physical conditions have been discussed by Roelfsema et al. (1996). Four

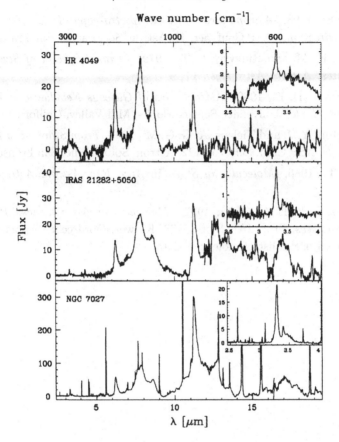

Fig. 4.3. Examples of PAH features observed in the post-AGB star HR 4049 and two planetary nebulae by the ISO satellite. The insets show the shorter-wavelength end in more detail. The background continua have been removed. Note that many fine-structure lines are also visible in the case of NGC 7027 (Beintema et al., 1996.)

different classes of PAH (UIB) spectra have been identified by Geballe (1997).

4.11 Further reading

Kaldeich, B. H., (ed.), 1988. *Infrared Spectroscopy in Astronomy, Proc. 22nd ESLAB Symp., Salamanca, Spain, 7–9 Dec., 1988.* ESA SP-290.

Kwok, S., 1993. *Astronomical Infrared Spectroscopy: Future Observational Directions.* ASP Conf. Ser. 41, Astron. Soc. Pacific, San Franscisco.

Merrill, K. M. and Ridgway, S. T., 1979. *Infrared Spectra of Stars,* in *Ann. Rev. Astron. Astrophys.,* **17**, 9.

Osterbrock, D. E., 1989. *Astrophysics of Gaseous Nebulae and Active Galactic Nuclei,* University Science Books, Mill Valley, California.

Pendleton, Y. J. and Tielens, A. G. G. M., 1997. *From Stardust to Planetesimals,* ASP Conf. Ser. Vol. 122, Astron. Soc. Pacific, San Franscisco.

Tsuji, T., 1986. *Molecules in Stars,* in *Ann. Rev. Astron. Astrophys.,* **24**, 89.

van Dishoeck, E. F., (ed.), 1997. *Molecules in Astrophysics: Probes and Processes,* Proc. IAU Symp. 178, Kluwer, Dordrecht. Appendix 2 contains a useful list of molecular databases.

5
Dust

5.1 Introduction

The central astronomical role of dust is at its most evident in the infrared. Protostars form from dusty clouds of molecular gas; the cool condensing dust around them emits copiously at sub-millimeter and far-infrared wavelengths. Even fully developed stars such as Vega may be found to be surrounded by remnant dust particles (see Habing et al., 1996), causing excess emission at long wavelengths. Near the end of their existence, a new generation of dust is formed by evolved stars, for example, in the atmospheres of asymptotic giant branch objects and in the ejecta of supernovae.

As in the visible region, dust scatters and absorbs light, giving rise to *extinction*, though its effects are much smaller in the infrared than in the visible. A dramatic example is the Galactic Center, which suffers 30 mag of extinction at V, so that only about one photon in 10^{12} comes through, whereas A_K (2.2 μm) is about 2.5 mag and $\sim$10% of the photons can penetrate. Figure 5.1 shows the distribution of several types of objects with cool dust in an IRAS color-color diagram.

Dust also polarizes the light from distant stars and some properties of the polarization are found to be related to the extinction.

Even when the extinction is quite moderate, by observing in the infrared, the effects of interstellar absorption, which bedevil the use of the cepheid and RR Lyrae period-luminosity relations for distance determination, may be greatly reduced (e.g., Laney and Stobie, 1993, in the case of cepheids).

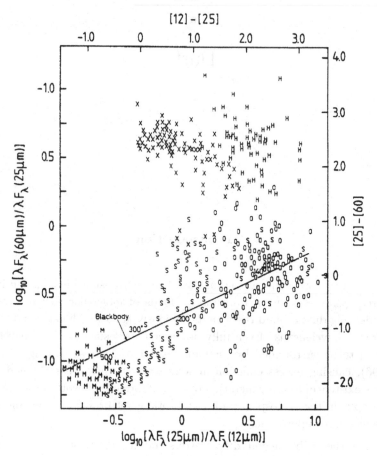

Fig. 5.1. IRAS two-color diagram from Pottasch (1993), showing a variety of objects whose SEDs (spectral energy distributions) are dominated by dust. The symbols are as follows: M, M stars and Mira variables; S, OH/IR stars; O, planetary nebulae; H, HII regions; X, galaxies. The blackbody line is shown for comparison.

5.2 Absorption and scattering by dust

The fractional reduction in intensity of starlight caused by dust grains is given by

$$\frac{\mathrm{d}I}{I} = -n_d C_{\mathrm{ext}} \mathrm{d}L,$$

where n_d is the number of grains per unit volume, C_{ext} is the *extinction cross-section* and $\mathrm{d}L$ is the path length.

The ratio

$$Q_{\text{ext}} = \frac{C_{\text{ext}}}{\pi a^2},$$

where a is the grain radius, is called the *extinction efficiency factor*. It is the ratio of the apparent cross-section of the dust particle to its geometric cross-section. It can be regarded as the sum of two terms, one due to scattering (Q_{sca}) and the other due to absorption (Q_{abs}). While a pure dielectric particle can only scatter light, a real dust particle both scatters and absorbs. The extinction – the effect it has on the light from background objects – is mainly determined by its scattering properties. On the other hand, the temperature of each particle is determined by its Q_{abs}, since this quantity controls how much it is heated by absorbing mainly high-energy quanta from the interstellar radiation field and how efficiently it can cool by emitting infrared photons. Each of Q_{abs} and Q_{sca} is dependent on λ/a and the complex refractive index of the grain, where only the real part is non-zero for pure dielectric substances. Absorbing substances have non-zero values for both real and imaginary parts. Ices and silicates may usually be regarded as almost pure dielectrics.

5.2.1 Scattering

It is found that scattering by interstellar dust particles can be modelled fairly satisfactorily by assuming that they are spherical, have a range of sizes and are mainly composed of dielectric substances.

Scattering theory indicates the existence of three distinct regimes, namely, when the typical particle radius a is much smaller than the wavelength, when it is comparable to the wavelength, and when it is much greater.

In the small particle limit, $a \ll \lambda$, it is evident that $Q_{\text{ext}} \propto \lambda^{-4}$, and the process is called *Rayleigh scattering* after Lord Rayleigh (1842–1919), who explained the blue color of the sky as being due to scattering of sunlight by small particles (air molecules). The scattered light is isotropic, i.e., has no favored direction, in this case.

For wavelengths within a small factor of a, the picture is more complicated, since resonances between the wavelength of the light and the size of the particle will occur, causing the value of Q to vary. The more complete Mie scattering theory (see van de Hulst, 1957) is then appropriate.

When $a \gg \lambda$, Q_{ext} is constant and the extinction is neutral.

Generally, the Mie theory is appropriate for interstellar scattering in the UV and visible, while the Rayleigh case describes the infrared. As a consequence, the particle size distribution does not affect the wavelength dependence of the infrared extinction very strongly. For this reason, there is much less variation in the infrared extinction law from one part of the sky to another than is the case in the visible and ultraviolet.

5.2.2 Absorption

The absorption properties of grains appear to be poorly known and rely mainly on theory. The *albedo*

$$\gamma = \frac{Q_{\text{sca}}}{Q_{\text{ext}}}$$

in the UV of the grains can be measured approximately. It normally shows a strong absorption feature at about $0.217\,\mu$m. In the infrared, the solid-state absorption features described in Chap. 4 may make their appearance, depending on the physical conditions of the interstellar medium through which the light must pass.

The continuous part of the absorption is thought to arise from the carbonaceous component of the dust. A deficiency of C in the inter-stellar medium towards the Galactic Center may thus be the reason for the anomalously low ratio $A_V/\tau_{9.7}$ seen in that direction (see section 4.10.1.1).

5.3 Practical aspects of interstellar absorption

For many purposes, the effects of interstellar absorption and the differential reddening effects it causes are something to be got rid of or compensated for. For example, the position of an object in a color-color or a color-magnitude diagram will be displaced to redward and made to appear fainter. In a given photometric system these displacements will be in a fixed direction given by the *reddening vector*. In section 5.3.1 the determination of the *reddening law* and related vectors are discussed.

5.3.1 R_V, the ratio of total to selective extinction

In the V or visible photometric broad band, the extinction in magnitudes is denoted by A_V. As the extinction is invariably greater in blue light than in the visible, it makes the dimmed object redder. The B-V color of the object is altered by an amount E_{B-V}, known as the *color excess* or *selective extinction*.

The quantity R_V, where

$$R_V = \frac{A_V}{E_{B-V}},$$

is called the *ratio of total to selective extinction*. For most directions in space, this quantity has a value close to 3.05. However, there are some lines of sight, particularly those passing through regions of recent star formation, where this ratio can be as high as 5, perhaps due to the formation of larger than normal grains through coagulation. The passage of starlight through dust also causes it to become polarized; the mechanism involves absorption by grains aligned by the interstellar magnetic fields. The wavelength at which the polarization reaches a maximum is found to be correlated with the value of R_V.

Frequently, the color excess is normalized to unit E_{B-V}, so that we have

$$\frac{E_{\lambda-V}}{E_{B-V}} = \frac{A_\lambda - A_V}{E_{B-V}} = R_V \left(\frac{A_\lambda}{A_V} - 1 \right).$$

In general, the extinction law applicable to all but the densest parts of the interstellar medium is found to be extremely uniform at infrared wavelengths.

It is only in dense molecular clouds that variations become apparent. In the infrared, they may be appreciable at the wavelengths of the solid-state absorption features (see section 4.10), such as those due to silicates at $9.7\,\mu\mathrm{m}$. Proceeding towards the visible, they become noticeable at around $0.9\,\mu\mathrm{m}$ and increase gradually towards the UV, where, of course, extinction variations are much more common.

It is expected that for very long wavelengths the extinction will tend to zero. However, Lutz et al. (1996) have found anomalously high extinction towards the Galactic Center in the 4–8 $\mu\mathrm{m}$ region compared to, for example, the Draine (1989) parameterization (described below).

5.4 Determination of the infrared extinction law

Extinction is generally determined by the *color-difference* method. The intrinsic colors or spectral irradiance of unreddened stars of a particular Morgan–Keenan type are assumed known, and photometric or spectro-photometric observations are made of stars of similar type which suffer extinction. From the differences, the wavelength dependence of the extinction can be derived.

In order to investigate the extinction at large distances within the Galaxy it is necessary to make observations of luminous stars. Especially for very early types, the number of these objects is small and the presence of circumstellar matter is fairly ubiquitous. It thus becomes difficult to define a sample of "clean" unreddened stars to use as standards. Usually, stars with significant circumstellar shells betray themselves by the presence of emission lines and can be avoided. Variable stars are also avoided if known. If the sample of a particular spectral type is big enough, its spectral energy distributions can be compared and the bluest of them taken to be unreddened. Obviously, population-dependent variations in the spectral energy distributions must be watched out for, not to mention the vagaries of particular astronomers who classify spectra.

When the method is applied in the infrared, the $B-V$ colors of moderately reddened standard stars are compared to the tabulated values for the appropriate spectral type and the infrared spectral irradiance is corrected according to an assumed law as a first approximation. The correct law is then derived from comparisons with more heavily reddened stars.

In broad-band work, such as $JHKL$ photometry, the precise details of the waveband passed by the filter affect the values of extinction obtained. It is desirable to be aware that most published work is to some extent heterogeneous because of the different filter sets and standard star compilations that were used. The more reliable measurements can be traced to particular instruments with published characteristics.

5.4.1 Extinction law in the near- and mid-IR

Several investigations of the interstellar extinction law in the $JHKL$ region have been carried out using the SAAO and other systems.

Generally speaking, the observed ratios are very close to those derived from the model of van de Hulst as long ago as 1946 and expressed as the "number 15" curve. This model assumed a rather arbitrary distribution of spherical dielectric particles designed to fit the visible data then available. The extinctions at various wavelengths were calculated according to Mie theory. In Table 5.1 the van de Hulst number 15 curve is compared with the work of Rieke and Lebofsky (1985).

The Rieke and Lebofsky (1985) results were obtained from a study of highly reddened cool stars near the Galactic Center, using the color-

Table 5.1. A_λ *values according to the van de Hulst no. 15 curve and the work of Rieke and Lebofsky (1985)*

Band	λ	A_λ vdH	R & L
U	0.36		
B	0.44		
V	0.55	1.00	1.00
R_C	0.64	0.78	
I_C	0.79	0.59	
J	1.21	0.245	0.282
H	1.65	0.142	0.175
K	2.2	0.081	0.112
L	3.45	0.036	0.058
M	4.8		0.023
N	10		0.052

Notes: λ is the wavelength of SAAO bands; vdH is a reading of the van de Hulst (1946) no. 15 curve; R & L is the law derived by Rieke and Lebofsky (1985) for stars towards the Galactic Center (late-type stars in or near the GC cluster).

difference method. Infrared spectra were used to classify the objects measured and their photometric colors were compared to normal, unreddened stars of the same spectral types.

A further determination was made by Landini et al. (1984), comparing theoretical to observed hydrogen-line ratios in the highly obscured HII region G333.6-0.2. The values they found for the extinction (taking $A_V = 1$) were about 10% less than those given by van de Hulst. They parameterized the extinction law as

$$\tau_\lambda \propto \lambda^{-1.85 \pm 0.05}.$$

The parameterization given by Draine (1989), $A_\lambda \propto \lambda^{-1.75}$, is also quoted frequently.

A general examination of the extinction law in the 1–5 μm region was made by Martin and Whittet (1990). Their favored power-law parameterization is

$$\frac{E_{\lambda-V}}{E_{B-V}} = 1.19\lambda^{-1.84} - 3.05,$$

appropriate to the diffuse interstellar medium.

The Rieke and Lebofsky (1985) curve implies greater extinction at infrared wavelengths and has a different exponent (-1.62) compared to the others.

While the ratios of the color excesses in various wavebands is fairly accurately known, the extinction for infinite wavelength is not, and has to rely on theory. It is generally assumed that

$$R_V = 1.1 \frac{E_{V-K}}{E_{B-V}}.$$

According to Whittet and van Breda (1978), this ratio is only weakly model-dependent for theoretical curves which fit the observations. By using this relation, the extinction at each wavelength can be found. It is important to realize that the absolute value of the extinction at K may thus be uncertain by as much as 10%.

5.4.2 Reddening vectors

When an object is reddened, its position in color-color diagrams will be displaced according to the appropriate *reddening vector*. The slopes of the vectors, such as E_{J-H}/E_{H-K} and E_{H-K}/E_{K-L}, and their lengths for unit A_V may be estimated from Table 5.1 or the parameterized relations mentioned in section 5.4.1. These values will depend on the precise effective wavelengths of the filters in use and may also change when the reddening is very high, since most of the photons they admit will be near the long-wavelength ends of their passbands. Further, if the intrinsic spectra of the objects of interest are peculiar, the color excess ratios will also be affected.

Table 5.2 lists some observationally determined color excess ratios for the SAAO (Carter, 1990) system, using the color-difference method (Glass, unpublished).

5.5 Polarization by dust

The grains forming the interstellar dust are believed in reality to be non-spherical and, as a result of processes such as paramagnetic relaxation, become aligned with the interstellar magnetic field. The scattering of light then depends on the direction of the electric vectors of the photons relative to the long axes of the grains, leading to linear polarization.

The percentage polarization in the visible region is seen to have a maximum at a wavelength which is related to R_V, the ratio of total to

Table 5.2. *Color excess ratios on the SAAO (Carter) system*

$\dfrac{E_{J-H}}{E_{B-V}}$	$\dfrac{E_{H-K}}{E_{B-V}}$	$\dfrac{E_{K-L}}{E_{B-V}}$	$\dfrac{E_{V-K}}{E_{B-V}}$	$\dfrac{E_{J-H}}{E_{H-K}}$	$\dfrac{E_{H-K}}{E_{K-L}}$	$\dfrac{A_V}{E_{B-V}}$
0.333	0.192	0.140	2.78	1.75	1.37	3.06
±0.007	±0.006	±0.013	±0.03	±0.07	±0.13	±0.03

Note: This is from a re-working of the Whittet and van Breda (1980) study of OB stars.

selective extinction. The cause of variation in the value of R_V relates to the sizes and shapes of the grains. The wavelength dependence of polarization can be fitted empirically in the visible to near-infrared regions, as the Wilking law (see Whittet, 1992):

$$\frac{P_\lambda}{P_{\max}} = \exp\left\{ -K \ln^2 \frac{\lambda_{\max}}{\lambda} \right\},$$

where

$$K = (1.66 \pm 0.09)\lambda_{\max} + (0.01 \pm 0.05).$$

This relation breaks down beyond $3\,\mu$m. It is found that, for $\lambda > 2.0\,\mu$m, the following law holds:

$$P_\lambda = P_1 \lambda^{-\alpha},$$

where P_1 is a constant for the line of sight and $\alpha = 1.8 \pm 0.2$. However, there are excursions above this relation which are associated with interstellar ice bands.

Polarization is also observed in the *emission* at far-infrared wavelengths from dust grains, and may be used to set some constraints on axial ratios if the grains are treated as spheroids (Hildebrand and Dragovan, 1995). The long-wavelength results tend to support the MRN model for grains (see section 5.6).

5.6 Models of dust grains

Dust models have to account for a number of known properties, such as sizes of the particles, their chemical compositions, extinction properties, polarization, absorption and emission characteristics. As observations have revealed more and more details of dust, the models have had to become more sophisticated. Not only must infrared observations be explained satisfactorily, but also the ultraviolet extinction (usually showing a strong feature at 2174 Å), the visible polarization properties and the interstellar absorption features.

The absorption caused by dust in the interstellar medium is surprisingly uniform and nearly linearly related to the hydrogen column density, enabling some general characteristics to be determined. For example,

$$\frac{N_H}{A_V} = 1.9 \times 10^{25}\,\mathrm{m^{-2}\,mag^{-1}}$$

where N_H is the column density of H atoms per m^2 in the interstellar medium and A_V is the extinction at V in mag. This result was obtained from Lyα and H$_2$ absorption line spectroscopy of stars within $\sim$1 kpc of the Sun (Bohlin, Savage and Drake, 1978).

The measurement of CO $J = 1$–0 and 2–1 in the millimeter wave regime is a more convenient means for investigating the density of interstellar gas. A useful general relation was derived by Young and Scoville (1982) for the 1–0 transition and modified by Maloney and Black (1988) as follows:

$$\frac{N(\mathrm{H_2})}{I(\mathrm{CO})} = 1.8 \times 10^{20}\,(\mathrm{K\,km\,s^{-1}})^{-1}.$$

The latter authors also discuss the likely effects of abundance variations on this ratio. The unit of CO intensity is the equivalent width of the CO line measured in terms of antenna temperature and Doppler-broadened line width. CO frequently becomes saturated (i.e., its optical depth becomes greater than 1) and its equivalent width is not then proportional to A_V. In such cases, use must be made of transitions of other lines such as the CO 2–1 transition or ^{13}CO 1–0, which remain linear to much higher densities.

The interstellar medium near the Galactic Center seems to contain relatively little dust for its CO emissivity (Glass, Catchpole and Whitelock, 1987). Other evidence for this effect comes from γ-ray studies.

As already mentioned, the early model of van de Hulst (1946) enjoyed a great deal of success in explaining the interstellar extinction curve. It was based on spherical dielectric grains with radii smaller than the wavelength of visible light. At the time, nothing was known of the detailed infrared extinction curve, but the model was quite successful in explaining its general character. Only with the discovery of individual "features" in the curve, around particular wavelengths (such as the silicate feature at 9.7 μm), did it become conspicuously inadequate.

A later widely accepted model is that put forward by Mathis, Rumpl and Nordsieck (MRN) in 1977. It explains the silicate bumps in the extinction curve as well as certain polarization properties. Only refractory grains of graphite and silicates were included. The size distribution varied according to radius to the power 3.3–3.6, between radii of 0.005 μm and 1 μm. This model was put on a more quantitative basis by Draine and Lee (1984), who calculated extinction efficiency factors (Q) for silicate and graphite grains.

5.6.1 Small grain regions

Certain zones of the interstellar medium show emission features (see section 4.10.3) which are usually ascribed to polycyclic aromatic hydrocarbons – the PAHs. These features are believed to be caused by situations where a strong ultraviolet source (such as a hot star) is in close proximity to a molecular cloud.

Sufficiently small grains, whether PAHs or not, may experience transient heating to high temperatures if they absorb even a single energetic photon (Aannestad and Kenyon, 1979). In this way, they may become much hotter (for a short while) than ordinary grains exposed to the same interstellar radiation field (see section 5.7.1). Such grains were proposed as the explanation for infrared emission from several reflection nebulae by Sellgren (1984).

5.6.2 Refractory dust grains

Grains all evaporate at some temperature, and the most refractory (the least easy to evaporate) are expected to consist of substances such as tungsten, corundum (Al_2O_3), perovskite ($CaTiO_3$) and graphite. The most refractory grains will be the first to form in the wind from a mass-losing star, at about 1500 K, and will be the last to survive in a region where destruction of grains by heating is taking place. Salpeter (1977) discusses the condensation possibilities for different regimes of temperature and pressure. Evidence for the formation of such grains comes from the fact that elements such as Al are under-abundant (depleted) in the interstellar medium compared to their overall cosmic abundance.

Carbon grains might be crystalline, like graphite and diamond, or amorphous, like soot. They may also be in the form of flat PAH plates, which resemble graphite.

5.6.3 Core-mantle grains

When the interstellar environment is cool and dense enough, more volatile compounds are able to condense. The refractory grains provide natural cores upon which H_2O, NH_3 and CH_4 can accumulate. In the densest molecular clouds, an outer layer of even more volatile substances, such as O_2, CO and N_2, may form. These composite grains are the likely site of the solid-state features observed in absorption (see section 4.10). The compositions of their mantles may be altered according to whether the H/H_2 ratio is small or large. If small, hydrogenation occurs, favoring production of H_2O, CH_4, NH_3 and CO_2. If large, CO and O_2 will form. The first category gives rise to "polar," or H_2O-rich, matrices and the second to non-polar. The formation of more complex organic compounds may be encouraged by UV bombardment (Sanford, Allamandola and Bernstein, 1997).

5.6.4 Large grains

At longer wavelengths, towards the sub-millimeter region, the Draine and Lee (1984) model predicts that the galactic emission is predominantly due to graphite. The absorption cross-section $Q_{ext} \propto \lambda^{-\beta}$, with $\beta \sim 2$. For amorphous carbon at these wavelengths, it is expected that $\beta \sim 1$, while silicates should lie between these values. There is evidence, however, that in circumstellar disks and some dense molecular clouds the opacity of the dust can be much higher than that predicted by the Draine and Lee model (Henning, Michel and Stognienko, 1995). This can be explained by the coagulation of dust to form much larger particles than are believed to exist in less sheltered parts of the interstellar medium.

5.7 Equilibrium temperatures of grains

5.7.1 Temperatures in the interstellar medium

Draine and Lee (1984) give the absorption efficiencies Q_{abs} for spherical graphite and "astronomical silicate" grains of different radii. For graphite particles of radius $0.05\,\mu m$, the UV absorption efficiency Q_ν is ~ 1. In the infrared, it is found that $Q_\nu = q_{IR}\nu^\gamma$, where q_{IR} is a constant (1.4×10^{-24}) and $\gamma = 1.6$.

The luminosity of one grain is $4\pi a^2 \pi Q_\nu B_\nu(T_{grain})$ W Hz^{-1}, where B_ν is the specific intensity of a blackbody.

For thermal equilibrium of a grain absorbing UV and radiating in the IR,

$$(\text{UV energy})_{\text{in}} = (\text{IR energy})_{\text{out}}.$$

In the low-density interstellar medium, although the energy distribution of the radiation might be characteristic of a high-temperature blackbody, its density is far lower than it would be in a blackbody cavity, and it is said to be *diluted*.

The radiation density formula

$$U = \int u_\nu \, d\nu = \frac{4\sigma}{c} T^4,$$

can be used in reverse, by Kirchhoff's law, to determine the temperature of a blackbody grain in a radiation field of known energy density. The value of U in interstellar space arising from stars is about $7 \times 10^{-14} \, \text{J m}^{-3}$, yielding $T \sim 3.1$ K, nearly the same as the cosmological background temperature. Here the dilution factor is around 10^{-14}. The radiation density from the 2.7 K cosmological background itself is undiluted and of similar density, but probably has little effect on grains because of their low emissivity (and hence absorptivity) at long wavelengths. In general, it is expected that the absorption efficiency Q has values

$$Q \propto \lambda^{-\beta},$$

where $\beta \sim 1$–2, in the far-infrared. In the cool (diffuse) interstellar medium the equilibrium temperature is typically 16 K (Greenberg and Li, 1996).

In a typical spiral galaxy, such as our own, the IRAS and 1.3 mm observations are satisfied by a two-component model with dust at 16 K and 50 K (Chini and Krügel, 1996). A very cold dust component could also exist, but would not lead to observable effects.

5.7.2 Temperatures near a strong UV source

Here we might consider the behavior of dust near an active galaxy nucleus. At a distance r from a powerful UV source of luminosity L_ν, the heating of a graphite grain (see section 5.7.1) by the incident UV flux is balanced by its infrared emission:

$$\frac{L_{\text{UV}} e^{-\tau_{\text{UV}}}}{4\pi r^2} \pi a^2 = 4\pi a^2 \pi q_{\text{IR}} \int \nu^\gamma B_\nu(T_{\text{grain}}) \, d\nu$$

or

$$\frac{L_{\mathrm{UV}}e^{-\tau_{\mathrm{UV}}}}{4\pi r^2} = 8\pi q_{\mathrm{IR}}\frac{h}{c^3}\left(\frac{k}{h}\right)^{4+\gamma} T_{\mathrm{grain}}^4\Gamma(4+\gamma)\zeta(4+\gamma),$$

where $e^{-\tau_{\mathrm{UV}}}$ is the UV absorption of the medium between the source and the grain, Γ is the gamma-function and ζ is the Riemann zeta-function. From this, for $\gamma = 1.6$, if grains evaporate at $T = 1500$ K, we can derive an expression for the minimum distance at which grains can exist without evaporation (where their temperature would be $\sim$1500 K):

$$r_{\mathrm{evap}} = 1.3L_{\mathrm{UV},39}^{0.5}T_{1500}^{-2.8}\,pc,$$

where L_{UV} is in units of 10^{39} W and T is in units of 1500 K. This formula, which is due to Barvainis (1987), has been applied to UV and IR observations of the very luminous Seyfert 1 galaxy, Fairall 9, by Clavel, Wamsteker and Glass (1989). The observed delay between changes in the UV flux from its nucleus and the response of its dust shell in the near-infrared implies, from the light propagation time, a distance which agrees with the predicted distance from the nucleus to the dust shell.

5.8 Life-cycle of the interstellar medium

Gas and dust are given off by the outer atmospheres of stars, especially post-AGB (asymptotic giant branch) objects, the long-period variables such as M- and C-type Miras and OH/IR stars. Supernova explosions play an important part when massive stars are plentiful. This material enters the interstellar medium, where it may form into dust and be processed and re-processed many times through the action of ultraviolet light, cosmic rays and other processes. Ultimately, the interstellar medium becomes the source material out of which new stars are formed. At the end of their lifetimes, these stars again return material to the interstellar medium, closing the cycle. However, the material becomes enriched in heavy elements through nucleosynthesis in the stars. The amount of material available decreases gradually as more and more of it is tied up in white dwarfs.

5.8.1 Mass loss from stars

Most of the mass loss which feeds the interstellar medium takes place in the late stages of stellar evolution of stars with masses of one to a few $M_\odot$. The AGB stars, especially the long-period variables, including the OH/IR sources, shed large fractions of their masses in winds of gas and dust, which cause them to be surrounded by envelopes of material

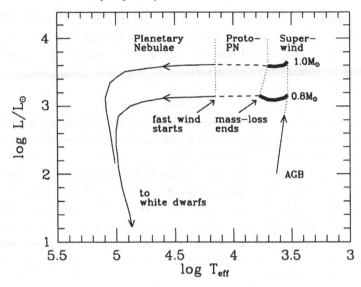

Fig. 5.2. Late-stage stellar evolution in the H–R diagram. The two tracks shown correspond to different AGB masses. Following heavy mass loss in the short-lived "super-wind" phase, the stellar atmosphere is reduced to $\sim 10^{-3} M_{\odot}$ and the core to about $0.6 M_{\odot}$, forming a proto-planetary nebula or PPN, surrounded and obscured by the ejected gas and dust. As evolution proceeds, the star becomes hot enough to ionize the ejecta and a planetary nebula is formed (based on a diagram from Schönberner, 1983).

at much higher density than the general interstellar medium. As they reach the ends of their lives, their mass-loss rates increase and give rise to a short-lived fast or "super-wind" phase (see Fig. 5.2) which may obscure the central star from visible or even near-infrared observation.

The superwind phase in turn merges into the PPN *(proto-planetary nebula)* phase. At this time, the central star is very luminous (class I; spectral type B–G) but has not yet become very hot. A PPN is distinguished by having a massive gaseous envelope showing molecular emission and a dust shell with temperature around 150–300 K. PPNs have been reviewed by Kwok (1993).

The planetary nebula evolves as the central star heats up and ionizes the remnants of the cast-off AGB envelope. As the nearer dust is evaporated, the average temperature of the remainder decreases due to the dilution of the UV radiation field with distance. The dust shell of the planetary nebula may show the 9.7 μm silicate or 11.2 μm SiC solid-state emission features which reveal whether the original AGB star was oxygen- or carbon-rich. In addition, young C-rich PPNs may show the

PAH features, even though these do not appear in AGB stars. They arise when a C-rich dense interstellar medium is bombarded by UV.

Eventually, the central star becomes a white dwarf, cooling to the point of invisibility over many Gyr.

5.8.1.1 Determination of mass loss

The current models for mass loss make use of simplifying assumptions, such as that its rate $\dot{M}$ is constant and that the geometries of the star and shell are spherically symmetrical.

An estimate of the total mass loss can be obtained from the CO component in the stellar wind alone. Its radial velocity and its angular extent are measured. The total emission (line width of ^{12}CO, or preferably of ^{13}CO on account of its lower density and consequent lower likelihood of being saturated) is used to estimate the mass of CO gas present and a ratio CO/H is assumed in order to derive the total outflow. This last step is particularly uncertain.

In another simple model involving the infrared emission from the dust within the outflow, it is again assumed that the mass-loss rate $\dot{M}$ and the outflow velocity are constants and that the shell is spherically symmetric, so that the density $n \propto R^{-2}$. Dust forms from the gas when it reaches about 1000 K, at a distance which is set by the dilution of the radiation field of the central star. Radiation pressure on the dust drives it outwards together with the gas, to which it is coupled by collisions.

The acceleration of the shell of dust and gas with thickness dr, density n m^{-3} and velocity v, at a radius r from a star of luminosity L, is given by the equation

$$nm\frac{dv}{dt} = Q(\lambda)n_d\sigma_d\frac{L}{4\pi r^2 c}$$

where m is the mean mass of a particle, and the quantities Q, n_d and σ_d are the absorption efficiency factor, density and cross-section of the dust particles, respectively (Pottasch, 1993).

The mass-loss rate can be written

$$\dot{M} = 4\pi r^2 vnm$$

and the optical depth of the dust

$$d\tau_d = Q(\lambda)n_d\sigma_d dr,$$

so that we can re-arrange the equations to give

$$\dot{M}\frac{dv}{d\tau_d} = \frac{L}{c}.$$

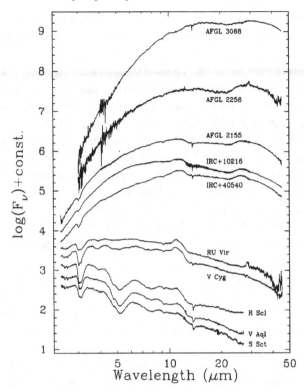

Fig. 5.3. A sequence of dust-dominated C-star spectra from ISO. The near-IR color temperature decreases from bottom to top. Note that apparent features at 9.3, 10.05, 11.05 and 28 μm in some stars are spurious. The broad SiC dust feature at 11 μm is seen in emission in the hotter objects and is self-absorbed in cooler ones, perhaps even appearing in absorption in AFGL 3068. The C_2H_2 feature around 3.1 μm and the fundamental CO at 5 μm become progressively weaker in the cooler stars. Around 14 μm are more C_2H_2 bands and also absorption arising from the C–H bending modes (Yamamura et al., 1998).

This equation may be integrated to yield (assuming that the initial velocity ∼0)

$$\dot{M} = \frac{L\tau_d}{v_t c} = 2 \times 10^{-3} \frac{\tau_d L}{v_t} \ M_\odot \ \text{yr}^{-1},$$

where v_t is the terminal velocity. The latter is estimated from the radial velocity of the CO ($J = 1$–0) line observed at mm wavelengths). The optical depth of the dust shell can be estimated from infrared spectral features such as the silicate band at 9.7 μm.

The agreement between mass-loss rates determined by this method and from CO alone is usually within a factor of 2, giving some reassurance.

Modelling of dust shells can be used to determine the dust mass alone and the total mass-loss rate can be derived by assuming a plausible gas-to-dust ratio. The innermost part of the dust shell, typically at about 6 stellar radii for a 2500 K star, will have a temperature of 1000 K, the point at which silicates are believed to condense. T_{dust} decreases with radius, and can be modelled by a radiative transfer program. Ideally, the predicted *spectral energy distribution* (SED) is compared over a large wavelength range, say 1–30 μm, with measured fluxes. Suitable measurements of late-type stars are starting to become available from the ISO satellite, but most of the available data cover only limited spectroscopic ranges (such as the spectra from IRAS) or simply consist of near- to mid-infrared photometry.

Schutte and Tielens (1989), for example, model an observed relationship between the *L–N* color temperature and the optical depth in the 9.7 μm silicate feature, using the mass-loss rate as a parameter. This feature is in emission at low mass-loss rates but goes into absorption for high ones.

More detailed models have been made by Justtanont and Tielens (1992), who take detailed account of the dust composition and particle size distribution. They model the temperature distribution of the dust with radius. As a result they obtain the mass-loss rates due to dust alone. In a later paper, Justtanont, Skinner and Tielens (1994) are able to compare, through further detailed modelling, the gas and dust mass-loss rates for several O-rich stars, obtaining typically

$$\frac{\dot{M}_d}{\dot{M}_g} \sim 5 \times 10^{-3}.$$

A review of methods for the determination of mass loss was given by Van der Veen and Olofsson (1990). Circumstellar envelopes and asymptotic giant branch stars have also been reviewed by Habing (1996).

A recent discussion of mass loss in carbon Mira variables has been given by Groenewegen et al. (1998). The ratio of SiC to amorphous carbon dust is a parameter of the model. The condensation temperature of these substances is likely to be 1500 K, compared to 1000 K for O-rich grains. Mass-loss rates in the long-period variables tend to be dependent on period and amplitude and are correlated with near- and mid-infrared colors (See Fig. 5.3 for a sequence of C-rich dust shell spectra).

Even more detailed modelling still needs to be carried out: the outer atmosphere of Mira itself has been studied by stellar interferometry at $\lambda = 10\,\mu m$ by Lopez et al. (1997). Although construction of a full-infrared image of Mira is not yet possible, enough is known about its spatial Fourier transform to deduce that its dust shell is in reality neither spherically symmetric nor constant in time.

In the Milky Way galaxy the rate of return of material to the interstellar medium from late-type stars and proto-planetary nebulae is estimated to be about 5 $M_\odot$ yr^{-1}. Supernovae in the Milky Way are relatively infrequent and contribute about 2 orders of magnitude less, though their detritus is rich in heavy elements. The dust component of the total mass returned is about 1/100 to 1/200 of the total.

5.9 Star formation

The removal of material from the interstellar medium is through the formation of stars. The study of the processes involved forms a major part of infrared astronomy. Protostars are believed to originate from dense clumps within cool molecular clouds. Their luminosity is derived at first from the gravitational energy associated with the collapsing material. The various stages of low-mass star formation, according to current beliefs, are outlined in Fig. 5.4. Classes I, II and III were originally defined by Lada (1987) and Class 0 by André, Ward-Thompson and Barsony (1993).

The earliest observable phase of star formation, the Class 0 (see Fig. 5.4) sources, are visible only in continuum radiation around 1 mm and in spectral lines of CO. It is believed that millimeter-wave continuum radiation is always optically thin, so that it can be used to estimate the mass of a cool dust shell. The Class 0 stars are so cool and their shells are so thick that they are not seen at $10\,\mu m$ in the mid-infrared or even by IRAS. However, there is evidence from centimeter-wave continuum radiation of collimated CO outflows from a central object. In other words, a hydrostatic core has formed. Matter is believed to be accreting onto this core in substantial amounts during the Class 0 phase, but the mass of material within the envelope still exceeds that in the "star."

Accretion continues into the Class I phase, where a protostar is believed to be almost complete, and is detectable at $2.2\,\mu m$. The accretion in Class I objects is still accompanied by collimated outflows of molecular gas, for example, CO. The amount of circumstellar material

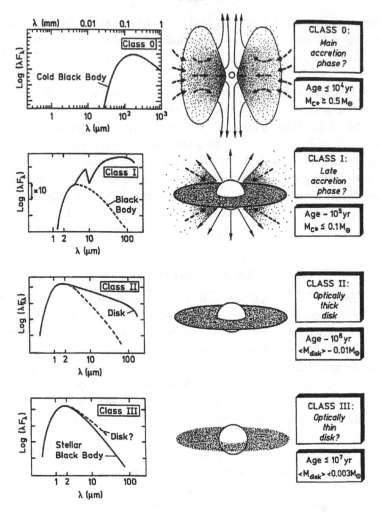

Fig. 5.4. Empirical evolutionary stages of different classes of YSOs (young stellar objects) according to André (1994). Class 0 represents the earliest stage of star formation, while Class III stars are close to being on the main sequence.

is substantial in a Class I object, perhaps as much as $0.1\,M_\odot$, and its 1.3 mm continuum radiation is noticeably extended.

Class II protostars – by this stage of development they are also known as YSOs (young stellar objects) – are much more compact in the millimeter-wave continuum, but still have thick accretion disks. CO outflows are no longer present.

Class III protostars have thin accretion disks and their spectral energy distributions approach those of normal stars. Classes II and III correspond to classical T Tauri stars.

The near-infrared spectra of YSOs are discussed by Greene and Lada (1996). Absorption line strengths increase with YSO class number, apparently due to a decrease in the amount of emission from luminous circumstellar material (*veiling*), which tends to obliterate them in the younger objects. Class I protostars show emission from H_2.

Evolutionary models for pre-main sequence (PMS) stars with masses up to $\sim$2.5 $M_\odot$ have been calculated by D'Antona and Mazzitelli (1994). They give the minimum mass for deuterium burning as $\sim$0.018 $M_\odot$ and, for lithium burning, as $\sim$2.5 $M_\odot$. The evolutionary tracks of the lower-mass models in the Herzsprung–Russell diagram are heavily dependent on the treatment of convection.

The circumstances which cause high-mass, rather than low-mass, star formation are not fully understood. Massive protostars are thought to manifest themselves first as ultra-compact HII regions, later becoming more obvious as luminous O or B stars in a normal HII region, to the excitation of which they contribute.

5.10 Dust and HII regions

The electron temperature within a classical HII region suggests that even the most refractory dust should not be able to exist. However, there is good evidence, especially for compact HII regions, that it does. The far-infrared flux that is measured correlates well with the number of Lyman continuum photons from the exciting stars (Jennings, 1975), implying that most of the stellar energy is ultimately absorbed by dust either within or on the periphery of the region.

For dust to exist in a stable condition in these circumstances, it must be shielded in some way from the intense ultraviolet radiation. Alternatively, it may be located in the photodissociation region (PDR) which is expected to surround a HII region in many cases. Real HII regions have complicated, non-spherically symmetric structures which are probably the consequence of density irregularities in the mother molecular cloud material.

5.11 Dust emission from the Milky Way galaxy

One of the most characteristic features of the IRAS maps at 60 and 100 μm are the wisp-like clouds called *infrared cirrus*. The shapes and locations of these are well correlated with other indicators of interstellar matter.

Data from the DIRBE experiment on the COBE satellite have been used to build up a detailed picture of the interstellar dust within 5° of the galactic plane (e.g., Sodroski et al., 1997). As will be remembered, this instrument covers the spectral region from 1 to 240 μm. By using pre-existing ^{12}CO, HI line maps and kinematic data as well as continuum information, the temperatures, masses and total luminosities of dust associated with the HI, HII and H$_2$ components of the interstellar medium have been established, for various galactocentric zones. The dust associated with the HI and H$_2$ components has a temperature of around 22 K, decreasing slightly with galactocentric distance from 2 to 10 kpc. The HII component has a typical temperature within the solar circle (i.e., for galactocentric distance less than 8 kpc) around 29 K, decreasing to around 22 K outside. The analysis shows that the interstellar radiation field increases towards the center. The PAH grains are relatively more abundant outside the solar circle than inside it.

5.12 Emission from dust in external galaxies

Imaging of spiral galaxies in the near-infrared often reveals surprising differences in their apparent morphology when compared to the visible region. This is a consequence of the interstellar extinction due to dust within the galaxies themselves and the fact that the extinction in magnitudes is reduced by a factor of about 10 at K- compared to the V-band (e.g., Block, 1996).

Most of the dust in galaxies is quite cool and its emission is only observable in the far-infrared. However, when there is starburst activity or an active nucleus present, there may be much hotter dust components, particularly in the nuclear region.

According to Chini and Krügel (1996), there is a direct relationship between the dust mass M_{dust} and the observed flux S_λ at $\lambda = 1300\,\mu$m, namely

$$M_{\text{dust}} = \frac{S_\lambda}{\kappa_\lambda B_\lambda(T_{\text{dust}})} D^2,$$

where κ_λ is the mass absorption coefficient of dust, estimated at 0.04 m^{2}kg^{-1}, T_{dust} is the dust temperature and D is the distance to the source.

An alternative way of determining the dust mass is through an assumed gas-to-dust ratio and the relation

$$M_{\text{gas}} = \beta L_{\text{CO}},$$

where β may have a value of 3–10. There is the expected degree of agreement between values for M_{dust} estimated in these two ways.

5.12.1 Normal galaxies

Normal spiral galaxies observed at IRAS wavelengths (12–100 μm) and longer show evidence for a cool dust ($\sim$20 K) component which peaks around 150 μm. The spectral energy distribution of this peak is broader than that of a single blackbody. With some assumptions about the nature of the dust, it can be deconvolved into components of two (or more) blackbodies at different temperatures. If the dust emissivity is assumed to vary with λ^{-2}, these have temperatures of 10–20 K (cold component) and 30–50 K (warm component).

Resolution-enhanced IRAS far-infrared (FIR; 40–120 μm) pictures of the disks of normal galaxies correlate very well with their images in Hα (e.g., Devereux, 1996). The histograms of log $L_{\text{IR}}/L_{\text{H}\alpha}$ show that their energetics must be very similar to those of galactic HII regions, so demonstrating that they are the origin of the warm dust component.

Towards the bulges of spiral galaxies there is also an enhancement of the 40–120 μm emission, but in this case the heating of the dust is believed to be due to the interstellar radiation field arising from the much greater overall density of stars.

5.12.2 Starburst galaxies

The distinction between a "normal galaxy" and a "starburst galaxy" is mainly a matter of degree. Further, it is not easy to distinguish between ongoing star formation activity and short-lived bursts.

Starburst galaxies show greatly enhanced mid-infrared emission thanks to the HII-region-like heating of interstellar material by large numbers of massive early-type stars. Their radio emission is dominated at first by free-free radiation but later is characterized by synchrotron emission from supernova remnants, as the massive stars come to the end of their relatively short lives.

Massive starbursts are usually spatially concentrated in regions surrounding the nuclei of their galaxies. The cause of the enhanced star formation remains an active subject of investigation. It is frequently associated with near-collisions of galaxies and actual mergers between them. It is also possible that the existence of a bar structure in a spiral enhances the build-up of density waves which encourage star formation by compressing molecular clouds to the point of instability.

Starbursts are known to coexist with Seyfert activity in some galaxies, such as in the bright Seyferts NGC 1068 (type 2) and NGC 7469 (type 1). As might be expected, the PAH features are conspicuous in starbursts while they only appear in Seyfert galaxies when there is other evidence for starburst activity. The expansion of supernova remnants gives rise to shock conditions in the interstellar medium which cause the excitation of characteristic features such as the [FeII] lines at 1.26 and 1.64 μm (see section 4.8.3.1).

The fine-structure lines arising from pure starburst galaxies are limited to those with ionization potential below 50 eV, whereas Seyfert galaxies show lines with much higher ionization potentials. Thus a diagram which shows the strength of a high-ionization-potential line against that of a PAH band is a good means for distinguishing starbursts from Seyferts (also see Fig. 5.5).

Even though starburst galaxies are highly gas-rich ($M_{\mathrm{gas}} \sim 2$–60×10^9 $M_\odot$), a high degree of starburst activity means that the available supply of gas and dust will be used up relatively rapidly. The mass consumption rate is estimated at 10^{10} $M_{\mathrm{gas}}/L_{\mathrm{IR}}$ (solar units) in years.

A recent review of starburst galaxies is that by Moorwood (1996).

5.12.2.1 Cosmological significance of starburst galaxies

The ISO satellite has been used to examine the Hubble deep field (HDF) at 6.7 and 15 μm, finding 15 objects identified with bright ($I < 23$) HDF galaxies. The detection rate is highest (2/3) for the irregulars, 1/3 for the spirals and lowest for nearby ellipticals (13%), suggesting on morphological grounds alone that luminous star-forming galaxies at moderately large redshifts preponderate (Mann et al., 1997). In addition, from ISO-based photometry, the SEDs of most of the detected galaxies show mid-infrared excesses, probably due to starburst activity (Rowan-Robinson et al., 1997) at rates of 8–1000 $M_\odot$ yr^{-1}. These results are in accord with results from deep surveys for star formation (HDF and others) which depend on UV emission (e.g., Madau et al., 1996; Connolly et al., 1997) and show increased star formation rates peaking at $z \sim 1.5$ (cf. section 3.9.5).

It would seem that, at very high z, star formation should only be manifested by the UV emission of hot stars and not by mid-infrared because the heavy elements from which dust is formed are not expected to be present in the first generation of stars. Another consequence should be the absence of interstellar extinction within primeval galaxies. However,

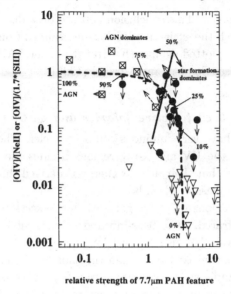

Fig. 5.5. Diagnostic diagram for nuclear vs. starburst activity. The vertical scale measures the ratio of high-excitation to low-excitation activity by means of fine-structure lines, and the horizontal scale measures the extent of the PAH feature, indicative of starburst activity. There is a clear progression from pure AGNs at top left to pure starbursts at bottom right (from Genzel et al., 1998).

some QSOs (see section 5.12.4), even though relatively young, appear to have substantial dust components.

5.12.3 ULIRGS

The *ultra-luminous infrared galaxies* (ULIRGS) represent the most extreme starbursts. They are defined by having $L_{IR} \geq 10^{12} L_{\odot}$. Their ubiquity in the sky was revealed only by the IRAS sky survey, as they are dust enshrouded and inconspicuous at visible wavelengths. They appear to be the result of strong interactions and mergers between gas-rich spiral galaxies.

Luminous infrared galaxies ($L_{IR} \geq 10^{11} L_{\odot}$) are more numerous than similarly energetic QSOs in the local universe ($z \leq 0.3$). Results from observations using the ISO satellite seem to be leading to a better understanding of their energy-production mechanisms. For example, Genzel et al. (1998) find, in a small sample, that about 75% of ULIRGS are powered predominantly by starbursts, while the remainder contain active galactic nuclei. This conclusion is based on a diagnostic diagram (Fig. 5.5) in which the excitation level as determined from infrared fine-structure lines is plotted against the strength of the PAH feature.

For a review, see Sanders and Mirabel (1996).

5.12.4 Active galaxies and QSOs

Active galaxies have very luminous compact nuclei. They are conspicuous objects in the UV and x-ray regions besides the infrared. The canonical view is that their nuclear energy is derived through accretion of matter onto a massive black hole.

The nuclear regions of active galaxies show excess radiation in the near- to mid-infrared. The near-infrared component is nearly always observed to be variable, usually over long time scales. Different explanations have been invoked to explain the shortest wavelength IR component; for example synchrotron radiation (e.g., Edelson and Malkan, 1986) and dust near its sublimation temperature of 1500 K (e.g., Clavel, Wamsteker and Glass, 1989; also see section 5.7.2).

In the outer parts of active galaxies, as the input of nuclear non-stellar energy increases, the IRAS and sub-millimeter fluxes are dominated by warmer components than are seen in normal galaxies. The warmer component is now typically 125 K and the cool one 33 K. The ratio L_{IR}/L_{gas} (solar units) becomes much higher. In normal spirals it is 5 ± 2, in Mkn galaxies it becomes 92 ± 53 and in radio-quiet QSOs it is about 550 (Fig. 5.6).

The spectral energy distributions (SEDs) of quasars from x-ray to radio energies have been studied by Sanders et al. (1989). The average quasar SED shows two "bumps," one in the UV and the other in the IR. The broad IR bump peaks around 3×10^{13} Hz or 10 μm, indicative of quite warm ($\sim$300 K) dust, if thermal. Somewhat less energy is radiated in the infrared bump than in the UV one. Sanders et al. consider that the mid- to far-infrared flux is generated by the disk of the host galaxy over a scale of several kpc. The disk must be warped if it is to allow adequate exposure of material to radiation from the active nucleus.

It is interesting to note that molecular gas and dust have been detected in some high-redshift QSOs, such as BR 1202–0725 at $z = 4.69$ (Omont et al., 1996). The continuum measurements were made at 1.35 mm, corresponding to a rest wavelength of 240 μm, and the CO(5–4) line was observed at 2.9 mm. This result shows that a copious metal-enriched interstellar medium already existed only 0.7 Gyr (for cosmological parameters $H_0 = 75$; $q_0 = 0.5$) after the Big Bang.

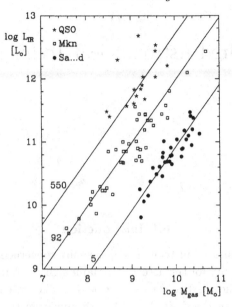

Fig. 5.6. L_{IR} vs. M_{gas} for normal spirals (•), active galaxies (open boxes) and QSOs (∗); taken from Chini and Krügel (1996).

5.13 Further reading

Bailey, M. E. and Williams, D. A., (eds.), 1988. *Dust in the Universe*, Cambridge University Press, Cambridge.

Bussoletti, E. and Vittone, A. A. (eds.), 1990. *Dusty Objects in the Universe*, Kluwer Academic Publishers, Dordrecht.

Cherchneff, I. and Millar, T. J., (eds.), 1997. *Dust and Molecules in Evolved Stars: UMIST Workshop, March, 1997. Astrophys Space Sci.*, **251**, pp. 1–495.

Dwek, E. and Arendt, R. G., 1992. *Dust-Gas Interactions and the Infrared Emissions from Hot Astrophysical Plasmas*, in *Ann. Rev. Astr. Astrophys*, **30**, 11.

Mennessier, M. O. and Omont, A., (eds.), 1990. *From Miras to Planetary Nebulae: Which Path for Stellar Evolution?*, Editions Frontières, Gif sur Yvette.

Pendleton, Y. J. and Tielens, A. G. G. M., (eds.), 1997. *From Stardust to Planetesimals*, ASP Conf. Ser. Vol. 122, Astron. Soc. Pacific, San Franscisco.

Whittet, D. C. B., 1992. *Dust in the Galactic Environment*, IOP Publishing, Bristol.

6

Infrared Astronomical Technology

6.1 Introduction

Astronomical infrared instruments are usually concerned with achieving maximum sensivity for the detection of faint signals. A major consideration is always how extraneous background radiation can be excluded. Only radiation of the desired wavelength and from the minimum possible area of sky should be seen by the detector, which is cooled to a low enough temperature for its own blackbody radiation and thermally excited background to be negligible. In addition, the detector is surrounded by a cavity at low temperature and the filter is cooled. Filters will radiate outside their passbands and even to some extent within them, as they are never wholly transparent.

In a photometer, the focal-plane aperture defining the part of the sky to be examined is cooled. The size of the aperture is chosen to be no larger than absolutely necessary, taking seeing conditions (image size) and the diffraction limit of the telescope into account.[†] This is one of the reasons why an array detector can give a better limiting magnitude than a photometer: in effect the aperture can be adjusted to the seeing conditions during data processing after the observation, and appropriate pixel weighting can be used to maximize the signal-to-noise ratio.

[†] The diffraction limit of a telescope's resolution is given conventionally (Rayleigh's criterion) by the formula

$$\theta_{\mathrm{arcsec}} = 0.252 \frac{\lambda(\mu\mathrm{m})}{a(m)},$$

where λ is the wavelength of interest in μm and a is the diameter of the telescope in meters. While at visible wavelengths the diffraction limit is rarely of concern, at $10\,\mu$m it is 1 arcsec for a 2.5 m telescope.

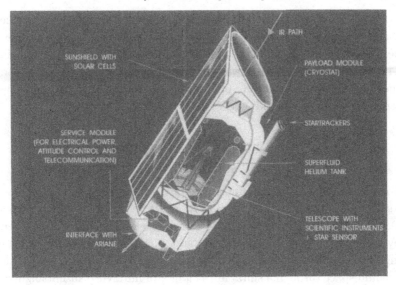

Fig. 6.1. Cut-away view of the ISO satellite, showing its helium-cooled Cassegrain telescope. This instrument was launched with the telescope under vacuum and at its operating temperature.

6.2 Infrared telescopes in space

In space it is possible to cool the entire telescope so that the background is determined only by celestial sources of radiation such as infrared cirrus and the zodiacal light (Fig. 6.1). If a liquid cryogen such as helium is needed, its boil-off rate is minimized by the use of multiple layers of radiation shielding, which may themselves be maintained at intermediate temperatures by heat exchange with the evaporating gas. The lifetime of a liquid-cooled infrared space telescope is clearly limited by its cryogen supply.

6.2.1 Passively cooled space telescopes

Using appropriate Sun shading, satellite orientation, surface finishes and multiple radiation shielding, it is possible to take advantage of the inherent frigidity of space itself to cool an infrared space telescope to very low temperatures without the use of cryogenic liquids. Such an instrument would have an indefinite lifetime, unlimited by supply of cryogen. A symposium, *Next Generation Infrared Space Observatory*, edited by Bell Burnell, Davies and Stobie (1992), was devoted to this subject.

SIRTF (also see section 2.8.3.2) will be launched warm and will make use of passive cooling to reduce the outer shell of the satellite to a temperature below 40 K, as well as to pre-cool the telescope following launch. The focal-plane instruments will be cooled by liquid helium to <2 K and the boil-off gas will be used to further cool the telescope to the operating temperature of around 5.5 K over a period of 40–60 days.

6.3 Construction components

6.3.1 Cryostats

To achieve the low background temperatures desired, the detector and its optics are placed within a *cryostat* or *Dewar* (named after Sir James Dewar, 1842–1923, inventor of the vacuum flask). Insulation is achieved by means of a vacuum jacket and "super-insulation," usually consisting of many layers of aluminized Mylar (Melinex). Radiation enters the vacuum space through a suitable "window" of infrared-transmitting material.

The materials used in the construction of unbaked metal Dewars outgas continuously with the consequence that the vacuum jacket will rather quickly cease to insulate effectively. This problem is solved by the use of *sorption pumps*, which contain materials, such as zeolites, that adsorb gases at low temperature and desorb them when warmed up.

While laboratory work is traditionally done using glass cryostats, infrared astronomy requires a more robust construction. Metal cryostats suitable for attachment to telescopes and capable of being tilted away from the vertical without losing the cryogen were developed by Frank Low around 1961. Experimental cryostats are frequently cooled by liquid nitrogen or liquid helium, which are transferred as required from insulated storage vessels.

Double cryostats are used when very low temperatures are desired. For example, the outer cryostat can be cooled by liquid nitrogen and the inner by liquid helium. The heat input to the inner vessel is much lower than that with a simple cryostat. Helium has a latent heat of vaporization of only 21 J g^{-1}, compared with 200 J g^{-1} for N_2. Without an outer cryostat, the hold time of liquid helium is very much reduced. Because of the trouble and expense of delivering these substances to remote observatories, more elaborate instruments often rely on *closed-cycle refrigerators*.

The temperature of a liquid cryogen can be lowered by pumping away the evolving gas. Liquid N_2 has its triple point at a pressure of 96 mm

Table 6.1. *Temperatures (K) of He and N_2 vs. vapor pressure (mm Hg)*

P	^{4}He	N_2
760	4.22	77.4
600	3.97	76.7
300	3.36	70.3
100	2.63	63.5
60	2.37	61.1
30	2.08	58.1
10	1.74	54.1
6	1.61	52.3
1	1.27	–

Note: 1 mm Hg corresponds to 133 Pascals (MKS units of pressure).

Hg and temperature 63 K. At this pressure, gas, liquid and solid co-exist. Below 96 mm Hg, N_2 is a solid and contact between it and the walls of its tank may become erratic, depending only on a few points of support, so that the temperature of the Dewar interior may start to rise or become unstable. This effect is countered by increasing the surface area exposed to the cryogen, through the use of copper swarf or other means.

Liquid helium has its λ-*point* at 37.8 mm Hg and $T = 2.172$ K. Below this temperature it is a superfluid. Table 6.1 shows the temperatures attained at various pressures.

6.3.2 Optical parts

Astronomical infrared instruments are somewhat more difficult to design than visible ones because they normally have to be operated at very low temperatures.

6.3.2.1 Refracting materials

Infrared-transmitting materials are needed for windows (for example, where an infrared beam enters a cryostat), filter substrates, detector substrates and lenses.

Because the wavelength range covered by the infrared is almost three decades, any given optical material is likely to be useful over a relatively small portion of it. While normal glasses become opaque around 2.2 μm, a few special ones work to 2.6 and 2.7 μm. A variety of crystalline substances transmit to much longer limits. These materials have high coefficients of thermal expansion and some are hygroscopic and/or soft and so may be difficult to work and handle. Some infrared-transmitting crystals are, in addition, birefringent.

In designing lenses to work at low temperatures, variation of refractive index with temperature as well as expansion coefficients have to be considered carefully, as does the differential expansion of the lenses relative to their holders. Some materials are liable to shatter if cooled too quickly. Accurate information concerning the variations of refractive index with temperature may be difficult to find or may not exist.

Some infrared materials have very high refractive indices n. For example, Ge has typically $n = 4.3$ and Si has $n \sim 3.4$. While these high-n materials have the advantage that the curvatures of the components need not be so great as for low-index substances, they require specialized anti-reflection coatings for good transmission. They are not transparent at visible wavelengths, complicating their optical alignment.

The *reflection loss* at the surface of a material of refractive index n is given, for normal incidence, by

$$R = \left(\frac{n-1}{n+1}\right)^2.$$

A surface of a material with refractive index n will give zero reflection loss (to air or vacuum) if it is coated with a $\lambda/4$ thick layer whose refractive index is $\sqrt{n}$. Multi-wavelength coatings can be devised to give good properties over a range of wavelengths.

Suitable pairs of materials for the construction of achromatic infrared doublets in the 1.0–2.5 μm range are given by Persson et al. (1992), who graph their partial dispersions and infrared "Abbé numbers." These authors also list the information available about the low-temperature properties of the common materials. The special, but very expensive, germanate Schott glass IRG2 can be used to make good achromats with BaF$_2$ and some other crystals. The Schott SF series can be combined successfully with BaF$_2$ or SrF$_2$ (Oliva and Gennari, 1995; 1998).

6.3.2.2 Reflecting components

Ideally, optical components should be designed as reflectors, which form images in the same way at all wavelengths and are easy to align using visible light, but suitable designs may not be available for applications requiring high resolution in combination with a large field of view. Gold

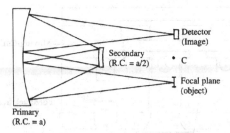

Fig. 6.2. The Offner (1975) relay. This system, which makes use of two spherical mirrors having the same center of curvature C, offers good low-distortion performance for 1:1 imaging. A cold stop can be introduced by placing baffles around the secondary.

is the preferred reflecting material in the infrared, and can be coated satisfactorily onto a glass substrate by using an intermediate layer of chromium. The chromium layer, which is difficult to remove subsequently, can be avoided by using an ion-assisted deposition process.

The best values of reflectivity and lowest emissivities, in the infrared, are obtained from films deposited under ultra-high vacuum (Bennett and Ashley, 1965). For example, gold at $10\,\mu$m has an emissivity of 0.006 when made with the UHV technique and 0.016 with a conventional system. It is essential that surfaces remain clean if their low emissivity is to be maintained. Of course, the emissivity of a mirror is less important if it is maintained at a very low temperature.

Spheric and aspheric mirrors are frequently constructed from solid aluminium by using numerically controlled machines having single-point diamond tools. These are capable of giving satisfactory finishes for infrared use thanks to the lower surface quality required at longer wavelengths. The surfaces are afterwards electroplated with gold.

6.3.2.3 The Offner relay

When 1:1 imaging is appropriate, use can be made of a very simple reflecting system with high performance discovered by Offner (1975). This "relay" and its variations feature in several infrared camera designs. It is, of course, achromatic. Figure 6.2 shows its basic layout. Both mirrors are spherical. By placing baffles around the secondary, a system of this kind can be used to introduce a cold stop into the optical train.

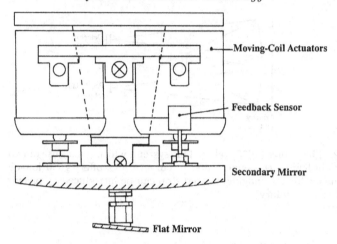

Fig. 6.3. Chopping secondary. The secondary mirror shown has a diameter of 18 cm and gives a Cassegrain focal ratio of $F50$ on a 1.9 m telescope. The mirror and the actuators are independently balanced and mounted on crossed-spring pivots. The flat mirror reflects a piece of de-focussed cold sky so that the reflection of the central hole of the primary is not seen by the detector. The dashed line outlines the support system (not shown).

6.3.2.4 Choppers

Chopping is frequently used to overcome fluctuations in the unavoidable background radiation (and random $1/F$ noise in the detector, if present). This technique involves subtraction of the signal from an empty piece of sky from that in the region of interest, usually at a frequency of several times per second. Chopping is accomplished typically by means of a stepwise-oscillating lightweight secondary mirror driven by actuators under feedback control (see Fig. 6.3). The need to minimize the time spent in movement between the two positions dictates that the moment of inertia of the mirror must be small, which means that it must have as small a size as possible and be made of lightweight material.

6.3.2.5 Telescope exit pupil and cold stops

The *exit pupil* of a telescope is the origin of the bundle of rays directed from its interior towards all usable points in its focal plane. In a prime-focus instrument or a Newtonian it is simply the primary mirror. In a Cassegrain, it is the image of the primary mirror in the secondary. Since this image is normally very close to the surface of the secondary, the secondary itself can be regarded as the exit pupil, i.e., as the "object" to be imaged by a pupil-imaging system such as a field lens.

A "Lyot stop" in infrared usage is a cold stop placed in the cryostat at an image of the telescope's exit pupil formed by a *field lens*. By the use of a suitable mask, all warm elements of the telescope within the pupil, such as mirror supports, can be hidden. Almost all infrared instruments incorporate such stops in their optical trains.

6.3.2.6 Fabry or field lenses

The Fabry lens is used in a photometer to image the exit pupil of the telescope onto the sensitive surface of the detector. By making sure that the image of the exit pupil of the telescope just fills the detector, the edge of the detector itself acts as a cold stop, and extraneous radiation, such as that arising from the telescope walls, is rejected.

The Fabry lens also has the important effect that it smooths out irregularities in the response of the detector as a function of position. Thus, a star merely needs to be within the measuring aperture and slight miscentering does not affect the response of the photometer. This effect may easily be visualized by considering that rays from a particular star or point on the sky will always be evenly spread over the surface of the detector, irrespective of how well the image is centered in the aperture.

6.3.3 Filters

Filters must pass only the wavelength band of interest and must exclude all others that the detector is sensitive to. In the 1–30 μm region they are usually constructed as multi-layer interference filters.

Interference filters are designed either to pass as narrow a band as possible, say to correspond to a particular spectral line, or to cover a range of wavelengths as uniformly as possible. Wavelengths outside the band are excluded partially by the interference properties of the filter and partly by the gross transmission properties of the coating and the substrate materials. The transmission of a broad-band interference filter within the passband is usually from 80% to 90%.

A useful introduction to infrared filter characteristics is given in Yen (1970). A less comprehensive wall chart was published by the same company (OCLI) in 1984. Macleod (1986) deals with the design and production of interference filters in detail. The items that must be mentioned in the specifications of a filter are listed in Table 6.2.

Table 6.2. *The specification of a filter*

Center wavelength
Full width at half-maximum
Transmission at maximum
Half-power points of the transmission curve
Slopes of cuton and cutoff
Transmission within passband (if a broad-band filter)
Permitted ripple (variation) within passband
Upper limit of transmission outside the band of interest
Variation of critical properties with temperature
Physical size: diameter, optically usable diameter
Optical quality (flatness)
Thickness and refractive index of substrate

The wavelength characteristics of an infrared interference filter change with departure of the direction of radiation from normal incidence. The transmission curve shifts to shorter wavelengths. Thus a narrow-band filter is degraded in sharpness and its peak transmission shifts to a shorter wavelength when the F ratio of an incoming beam is made large. The shifts can amount to a few percent. For example, at 20° the shift is between 0.5 and 1% and at 40° it is from 1.8 to 3.2% (Yen, 1970). Filters are occasionally placed at a small angle to the optical axis of an instrument in order to avoid unwanted reflections or "ghosts."

When the temperature of operation becomes lower, the transmission curve also shifts to shorter wavelengths. The shifts can amount to 2 or 3% between ambient and cryogenic temperatures.

Circular variable filters (CVFs) are a form of interference filter in which the thickness of the layers increases as a linear function of angle. They can be used to construct simple spectrometers with resolutions of $\Delta\lambda/\lambda \sim 1\%$. A single filter segment can cover as much as a factor of 2 in wavelength.

Where a narrow-band, tunable, filter is required, an electrically controllable Fabry–Perot etalon can be used, with an interference filter as a pre-filter or order-sorter.

6.3.3.1 Filters for $\lambda > 30\,\mu m$

At wavelengths beyond 30 μm, the layers of dielectric interference filters have to be quite thick and the effects of differential thermal expansion can no longer be tolerated. Instead, broad-band filters make use of intrinsic and extrinsic absorption bands in bulk crystalline materials, organic films, etc., as well as scattering by diamond particles of appropriate size.

For example, the 100 μm filter in IRAS (Beichman et al., 1988) is made up as follows: for short-wave blocking, an interference filter, followed by layers of sapphire, CaF_2, KCl and diamond powder; for defining the short-wave cuton, KCl; and for the long-wave cutoff, Ge doped with Ga.

Finding suitable combinations of materials for constructing filters in this manner is clearly something of an art.

More specific filters can be made by depositing metallic patterns such as dots and grids on a transmitting substrate to give appropriate cuton and cutoff wavelengths.

6.3.4 Polarizing components

Tinbergen (1996) gives a comprehensive account of astronomical polarimetry, including some information on infrared instruments.

In the *JHK* region, achromatic retarders of MgF_2 are available as well as Rochon prisms of LiNbO (Hough, Peacock and Bailey, 1991). At longer wavelengths, CdS retarders and wire-grid polarizers are used (Smith, Aitken and Moore, 1994).

An imaging polarimeter used in connection with a HgCdTe array has been described by Hough, Chrysostomou and Bailey (1994). This instrument has a MgF_2 Wollaston prism analyzer and a rotating waveplate. It enables the e- and o-rays to be observed simultaneously, reducing effects caused by variability of atmospheric transmission. The field is sliced into parallel strips by an input mask, so that half of it can be observed at a time. A similar instrument is used on UKIRT.

6.4 Individual detectors

As mentioned in Chapter 1, the infrared starts to be considered different from the visible on technical grounds at a wavelength of 1.1 μm. This represented the limit at which individual photons could be detected, at least until recently. Typical detectors are classified as *photovoltaic* or *photoconductive*, according to whether they register photons by generating a current of electrons or merely by changing their resistance.

Detectors are made of semiconductors and rely on the excitation of electrons from an energy band in which they are immobile to the conduction band, where they are free to move about. The difference in energy levels determines the minimum energy required of the incident

Table 6.3. *Maximum usable wavelengths for several common detector materials*

Material	Temp (K)	λ_{cutoff}
Si	295	1.11
Ge	295	1.85
InSb	77	5.4
HgCdTe[1]	77	2.5
Si:As	5	23
Si:As[2] (BIB)[4]	5	30
Si:Sb		36
Si:Sb[2] (BIB)[4]	5	40
Si:Ga[3]	10	17.5
Ge:Ga	–	115
Ge:Ga (stressed)	–	>200

Notes: 1. By changing the detailed composition, the bandgap of HgCdTe can be adjusted over a considerable range. The figures shown here are for the NICMOS chips. 2. This is from Stapelbroek et al. (1995). 3. This is from Lucas et al. (1995). 4. See section 6.4.3.

photon if it is to have the desired effect. Materials which have suitable natural band structure are called *intrinsic*, and those which are doped to provide suitable levels or bands are called *extrinsic*. The maximum wavelength that can be detected is given by the formula

$$\lambda_{\text{cutoff}} \ (\mu m) = 1.24/E_{\text{excit}},$$

where E_{excit} is the excitation energy in electron volts (eV).

The characteristics of a few common detector materials are given in Table 6.3.

The smaller the bandgap, the more likely electrons are to enter the conduction band unwanted, by thermal excitation. It is therefore necessary to cool these longer-wavelength detectors more than the shorter-wavelength ones.

Detectors are characterized by their quantum efficiency, linearity, response time and dark current.

The *quantum efficiency* is the fraction of the incident photons that produce collectable conduction electrons in the detector.

The *linearity* of a detector must be checked, for example, by observing standard stars of widely different magnitudes. Compensation must be applied if non-linear behavior is found.

The *response time* of a detector to a change in photon flux is also an important characteristic, affecting readout speed and frequency response.

The *dark current* is a source of noise. It consists of spurious conduction electrons arising from sources other than photon detections, such as thermal excitation and electrical leakage within the array.

The *readout noise* is usually expressed as the error in determining the number of electrons collected during the period of exposure. It is more a property of the readout electronics than of the detector itself. For example, if an array has a readout noise of 30 electrons, it would be exceeded by the Poisson statistical fluctuations in the detected signal only if $\sqrt{n_e}$ exceeds 30, where n_e is the number of conduction electrons.

Recent developments in detectors have included a *solid-state photomultiplier* which offers single-electron detection in the near- to mid-infrared. The principles of these devices were discussed by Rieke (1994). Hayes et al. (1989) give some details of performance, such as a quantum efficiency of order 20% at $10\,\mu$m.

6.4.1 Photoconductors

A photoconductor, as its name implies, is a material which increases its conductivity when illuminated. From a noise point of view, it suffers the disadvantage that a current must be passed through it to detect the change in resistance produced. A common effect of this is noise with a power spectrum that varies according to $1/F$, where F is the measuring frequency. This type of noise is called "flicker" or $1/F$ noise. Its extent is kept small by careful choice of contact type and bonding technique. It appears to arise where surfaces contact each other, for example, between the very small crystals that make up a chemically deposited PbS detector.

Photoconductors are usually made of doped Si or Ge. The dopant is chosen to provide impurity levels or bands at a level below the conduction band suitable for a particular range of infrared photon energies. Detectors of this kind are favored for wavelengths beyond $5\,\mu$m. The choice of dopant also depends on the absorptivity of the compound at the wavelength of interest and the thermal dark current generated.

Photoconductive gain relates the number of electrons collected at the output of the device to the number of photons giving rise to conduction electrons. It is typically less than 1 due to loss of electrons to recombination before they are collected. However, in some devices, such as the solid-state photomultiplier and some BIB detectors, gains in excess of 1 may be obtained, though often with the introduction of extra noise.

6.4.2 Photodiodes

Photodiodes (photo-voltaic devices) generate a current proportional to the number of incident photons. They are operated where possible at zero voltage, in a feedback circuit, to avoid $1/F$ noise. Generally, single-element detectors are limited by the input noise characteristics of the field-effect transistors that amplify their signals.

Much better noise performance is usually obtained from modern array detectors than from single-element devices, essentially because the individual elements are physically small and so have small electrical capacity, making for a higher output voltage per electron, which overcomes the readout noise.

6.4.3 Blocked impurity band (BIB) detectors

When it is attempted to make heavily doped extrinsic photoconductors (in order to obtain high quantum efficiency), it is found that the resistivity of the material becomes too low for the construction of a low-noise detector in the normal way. For this reason, an insulating layer of pure undoped material is introduced between the detector layer and the metallic contacts on the back of the device which are used to connect it electrically to its readout circuit. Such a layer allows the movement of charges promoted into the conduction band but blocks charge migration in the impurity band. (The other connection of each detector is made to an optically thin buried conducting layer on the input side.)

The same principle applies in the case of IBC (impurity band conduction) detectors. The names BIB and IBC are used by Boeing (formerly Rockwell) and Raytheon (formerly Santa Barbara Research Center) for detectors constructed according to the same principle.

6.4.4 Bolometers

An ideal *bolometer*, by definition, is a device that detects all the radiation falling on it. Although various forms of bolometers have existed for 100 years or more, astronomical interest is restricted to their modern form, in particular the Ga-doped Ge bolometer developed by Low (1961) and doped Si devices operating on similar principles. These devices are neither photoconductors nor photodiodes. A typical detector consists of a small chip of the doped material supported by very thin wires which act as electrical conductors for the measurement of its resistance and at the same time connect it to a heat sink with a certain thermal resistance, which has to be chosen in advance according to the background level of radiation that is expected to strike it. The doping level of the material is chosen to provide an optimum sensitivity of resistance to temperature at around its operating temperature, which is typically 1–2 K. The surface of the detector is blackened with a suitable absorptive paint. The sensitivity of the device is wavelength independent, as long as the paint is absorptive and as long as the dimensions of the detector are larger than the wavelength of the radiation.

The coefficient of change of resistance with temperature of a Ga-doped Ge bolometer is dependent on its operating temperature, which means that it also depends on the background flux striking it. Since this may vary according to the filters in use, as well as ambient conditions, care must be taken to establish its sensitivity with sufficient frequency and for each waveband.

6.5 Detector arrays

A multi-element array consists of a large number of picture elements or *pixels* arranged in rows and columns.

Whereas efficient visible arrays can be made on silicon and so are able to take advantage of mature manufacturing technology, the same is not quite true for the infrared.

Infrared arrays are usually made at present as two-layered devices. The upper layer consists of a suitable infrared-sensitive material formed into photoconductors or photodiodes. The readout of the array is done through the bottom layer, called the *multiplexer*, which is constructed using conventional Si-based techniques. The performance of the multiplexer is in many ways as critical to the success of the array as that of the infrared-sensitive layer. The two parts are joined electrically with one connection for each pixel. The interconnection is achieved through the use of small indium pillars called "bumps." It is the ideal material for this purpose because of its malleability and low melting point. During manufacture, the two layers are aligned under an infrared-viewing microscope and the bumps are welded by pressing them together.

The layers of an infrared array inevitably have different thermal expansion properties, and repeated cooling cycles may lead to difficulties such as detachment of the bump bonds.

Each individual photodiode and its associated electronics (located on the silicon part of the device) possess a certain electrical capacitance C. Normally the voltage V across this capacitor is set to a particular value (in the "reverse" or non-conducting direction of the diode) when the pixel is "reset" at the beginning of an exposure. Electrons released as a result of incoming photons discharge the capacitor according to the usual relation

$$\Delta V = \frac{\Delta Q}{C} = \frac{n_e e}{C},$$

where n_e is the number of electrons and e is the charge on an individual electron (1.6×10^{-19} coulomb).

Unfortunately, the capacitance of a diode is dependent on the voltage across it, so that the relationship between voltage and number of photons is to some extent non-linear, requiring compensation during image processing. A discussion of this problem for the NICMOS 3 array is given by Luginbuhl et al. (1995).

The lower the capacitance, the higher the voltage developed for each photon, and the easier it is to overcome the inevitable electronic background noise. However, the capacitance can only accept a certain number of electrons before it is fully discharged, and, under conditions of high background, it is desirable that this quantity, known as the *well depth*, be as large as possible in order to permit reasonably long times between readouts. Thus the requirements for high- and low-background conditions differ.

The design of the multiplexer electronics for long-wavelength detectors is very much related to the background conditions that are expected. Different systems are used when the background is low, such as in space-borne cryogenic telescopes. Ground-based broad-band cameras have high backgrounds, whereas ground-based spectrometers represent an intermediate case.

The duration of the exposure in an infrared camera is not controlled by a shutter because it would have to be cold to avoid radiating and also because it would usually have to be rapid acting. Instead, the exposure time of a pixel is effectively the time between reset and readout. One position of the (cold) filter wheel is usually made opaque to enable measurements of the array to be made with no radiation falling upon it.

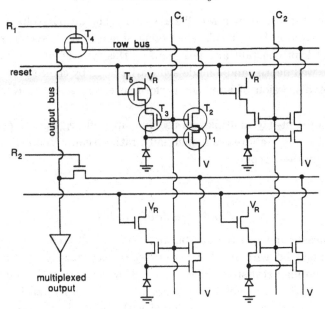

Fig. 6.4. The circuitry around four pixels of a direct readout array. The detectors are represented by diode symbols and are buffered by source followers. There are four transistors for each pixel. The column select signal C and the row select R are generated from the x and y shift registers. When both are present, the detector output is fed to the output bus. If the reset signal is present, the detector voltage is reset (taken from Rieke, 1994, by permission).

6.5.1 Readout electronics

The readout of most infrared arrays is called *direct*, because the voltage levels of the pixels are read out directly, one by one, unlike in a visible-region CCD (charge-coupled device), where the charge is first fed to a resettable integrating circuit. In a direct array, a pixel to be read out is "addressed" by its x and y coordinates, and the voltage on the pixel, buffered by its private source follower, is connected directly to the output bus of the chip (Fig. 6.4). Each pixel also has a MOSFET switch which can be used to connect it to the reset bus, if resetting is desired.

To avoid the problem of non-linearity caused by storing charge within the detector, a variant on the standard readout circuitry has been developed, incorporating a *capacitative trans-impedance amplifier* (CTIA). Each detector is connected to a high-gain DC amplifier with capacitative feedback. The photocurrent is collected in the feedback capacitor C

until it is discharged by a MOSFET switch. The output voltage of the stage is given by $V = n_e/C$. The potential across the detector is automatically kept constant and near zero by this type of circuit. In addition, the well depth can be made arbitrarily large by choice of the value of C. A discussion of amplifier performance was given by Kozlowski (1996).

Most present-day multiplexers use on-chip shift registers to generate the pixel x and y addresses sequentially, rather than individual address lines or address decoders.

6.5.2 Readout modes

Every measurement of the number of electrons collected on a pixel has noise associated with it. If the number of electrons is n_e, the noise from Poissonian statistics alone is $\sqrt{n_e}$. However, when the number of electrons detected is small, the Poissonian noise may often be smaller than the rms *readout noise* associated with the readout process of the array itself. This latter type of noise, as for a single detector, is usually expressed in terms of equivalent numbers of electrons. In what follows, it will be seen that noise may be minimized by attention to the readout procedure employed.

6.5.2.1 Noise reduction

Because a direct readout array can be read many times without affecting the charges accumulated on the individual detectors, an improvement in the random readout noise can often be obtained by making multiple measurements and averaging them. If the Poissonian noise of the collected electrons should be dominant, naturally no improvement will occur. In arrays which show glow from their output amplifiers, multiple readouts may actually increase the Poissonian noise by increasing the background.

6.5.2.2 Simple readout

In this mode, an exposure is made by simply resetting each pixel, exposing, and reading the voltage developed at the end of the exposure time. Unfortunately, this method is very noisy due to uncertainty in the voltage level at the start of the exposure, arising from a fundamental limit to the accuracy of the reset process called *kTC noise*.

6.5.2.3 Correlated double sampling

Correlated double sampling involves making two measurements: one directly before a change being measured and the other directly afterwards. By taking the difference, the effects of (a) the uncertainty in the initial voltage level and (b) the long-term (i.e., low-frequency) drifts are reduced. This technique is used very effectively in reading out the charge packets from CCDs (charge-coupled devices such as visible-region arrays and some early infrared arrays).

Correlated double sampling may be used with direct readout arrays, by making the first measurement at the time a reset is performed or directly afterwards. The second measurement is made at the end of the exposure, just before the next reset.

6.5.2.4 Reset–read–read

In this improved version of correlated double sampling, all pixels of the array are reset in sequence before the beginning of an exposure, without reading the output voltages (see, e.g., Fowler and Gatley, 1990). The completed reset is followed by the first readout sequence. The array is read out again at the end of the exposure time. The desired signal is the difference between the two reads. This is the basic procedure in use at the present time.

6.5.2.5 Multiple (Fowler) sampling

By reading the whole array n times each at the beginning and end of the exposure, and averaging before subtraction, the noise may be reduced by a factor of order $\sqrt{n}$. It is important to note that each pixel must be addressed freshly each time it is read: the noise reduction does not occur if the signal is merely digitized n times. The source of the noise is related to the addressing of the pixel, though its cause is not well understood. The signal-to-noise ratio produced by multiple sampling has been examined by Fowler and Gatley (1990) and Garnett and Forrest (1993). They show that, in the read-noise limited case, optimal results are obtained by sampling continuously for the first and last thirds of the exposure but omitting the middle third.

6.5.2.6 Continuous readout

It is possible to optimize the readout still further by the technique of *continuous readout*, where the readout process commences at the start of the exposure and continues (without reset) until its completion. The output voltage is closely proportional to the integration time and the slope of its value vs. time becomes the measured quantity. Each measurement effectively reduces the error in slope, so that the readout noise experienced in ordinary correlated double sampling is again reduced.

Garnett and Forrest (1993) also show, again for the read-noise limited case, that this line-fitting process is slightly superior to optimal multiple Fowler sampling.

6.5.3 Array controllers

The controller is an electronic box which provides a variety of well-stabilized fixed voltage levels and pulsed waveforms suitable for driving the x–y addressing elements and reset circuitry of the array. In addition, it controls the operation of one or more sampling voltmeters used to measure the output signals. The latter depend on *analog-to-digital converters*, which usually have a resolution of 16 binary bits (65,536 discrete levels). The speed with which the converters can function determines the time taken to read out the array, together with the bandwidth of the multiplexer electronics and the transfer characteristics of the signal train. It is always desirable to keep these as short as possible in the interests of efficiency, but it becomes essential when the background is high and the array has to be read out many times each second.

6.5.4 InSb arrays

The leading manufacturer of InSb arrays suitable for astronomy is Santa Barbara Research Center (now called Raytheon Infrared Center of Excellence). Typically, the 256×256 array is in use at the present time. The 1024×1024 array (known as the ALADDIN) is still under development (Table 6.4).

To achieve good broad-band performance at the longer wavelengths, where the background is high, the arrays have to be read out many times each second. The multiplexer of the 1024×1024 array has 32 separate outputs, requiring 32 preamplifiers and 32 analog-to-digital converters in order to improve the readout speed. A detailed description of the array has been given by Fowler et al. (1995).

A performance report on the 256×256 array has been given by Fowler et al. (1994), and a comparison of two types of SBRC array with one from Cinncinnati has been published by Finger et al. (1995).

Table 6.4. *SBRC (Raytheon) InSb arrays used for astronomy*

Format	Pitch (μm)	Well Depth Electrons	Noise Electrons (rms)	QE
58×62	75	3.2×10^5	200	80%
256×256	30	2×10^5	10–50	$\geq$80%
1024×1024	27.5	3×10^5	25	80%

Notes: The fill factor (sensitive area as fraction of total area) is $\geq$90%. These arrays are operated at a temperature of 30 K. The operability (percentage of good pixels) is $\geq$99%.

6.5.5 HgCdTe arrays

The most commonly used array detector in infrared astronomy is the NICMOS 3, developed for the Hubble Space Telescope NICMOS instrument by Rockwell International Science Center (now part of Boeing). Its HgCdTe layer is formed on a robust sapphire substrate, through which the incident radiation must pass. This technique offers fewer thermal expansion problems than CdTe substrates used previously, which were prone to cracking during the bump-bonding process.

This technology has been extended successfully to arrays of dimensions 1024×1024 (e.g., the so-called "Hawaii array"), and devices of four times this size are under development (Table 6.5).

6.5.6 PtSi arrays

Platinum silicide offers the advantage as a detector material that it can be made relatively simply as part of a normal silicon chip. For this reason, large area arrays can be constructed with excellent cosmetic properties. They have the drawback, however, of low quantum efficiency (QE), amounting to <10% at 1–3 μm, exacerbated by the need to place readout circuitry between the photodiodes. A camera based on a Mitsubishi monolithic PtSi array with 1040×1040 pixels was described by Glass, Sekiguchi and Nakada (1995). PtSi arrays are also made with near-100% fill factor by using sandwich-type construction.

Table 6.5. *Boeing (formerly Rockwell) HgCdTe arrays used for astronomy*

Format	Pitch (μm)	Well Depth Electrons	Noise Electrons (rms)	QE
256×256[1]	40	10^6	<40	55–70%
1024×1024[2,3]	18.5	1.2×10^5	3	60–70%

Notes: 1. The NICMOS 3 array has been superseded by the PICNIC device, which has a slightly different multiplexer. 2. Hodapp et al. (1995) state that the fill factor (sensitive area as fraction of total area) is $\geq$90%. These arrays operate at about 77 K, the temperature of liquid nitrogen. The operability (percentage of good pixels) is $\geq$99%. 3. A. F. M. Moorwood (private communication) states that the noise figure was obtained by multiple nondestructive sampling in a 1 min exposure. The dark current was $\sim$20 $e^- h^{-1}$.

6.5.7 Mid-infrared arrays

The problems associated with the very high background experienced when mid-infrared arrays are used with ground-based telescopes have been outlined by Gezari et al. (1992). For example, the thermal background flux at the Cassegrain focus of a large telescope is typically 10^9 photons s^{-1} m^{-2} μm^{-1} arcsec^2 around 10 μm. As the well capacity of a typical detector may only be of order 10^5 – 10^6 electrons, although well capacities of up to 3×10^7 electrons are available in some arrays, speed of readout is a serious problem. In the instrument described, pixels were 0.26 arcsec^2, the efficiency of the optics was $\sim$0.5, the photoconductive gain of the detector was $\sim$0.1 and the integration time was $\sim$0.03 s.

The camera described by Gezari et al. (1992) is based on a 58×62 element Si:Ga photoconductor array by SBRC (Raytheon), with direct readout. When read out 30 times per second it is capable of detecting a 0.1 Jy source (assumed spread over 25 pixels) at 1σ level in 1 min, with a 3 m telescope and a filter bandwidth of $\sim$1 μm.

Because of fluctuations in the atmospheric emission, which have a $1/F$ characteristic frequency dependence, it is necessary to use a chopping secondary, operating at a frequency of a few Hz. Even so, there is some residual background after subtraction of the images, because

Table 6.6. *Some mid-infrared arrays for IR astronomy*

Material	Format	Company	Reference
Si:As IBC	320×240	SBRC (Raytheon)	Venzon et al. (1995)
Si:As BIB	256×256	Boeing (Rockwell)	
Si:As BIB	128×128	Boeing (Rockwell)	Stapelbroeck et al. (1995)
Si:Ga	128×192	CEA-LETI-LIR	Lucas et al. (1995)

the detector "sees" a slightly different view of the telescope in the two chopper positions.

Table 6.6 gives some information on other mid-infrared arrays, developed or under development.

6.5.8 Far-infrared arrays

The SIRTF home page lists two far-infrared arrays under development for its MIPS instrument. They are a Ge:Ga 32×32 array operating to $70\,\mu$m and a stressed Ge:Ga array of 2×20 elements for use to $160\,\mu$m.

6.5.9 Handling of arrays

Infrared array detectors are exceedingly expensive even when compared to visible CCDs thanks to their experimental nature and the small numbers of them that are produced. They are also very liable to damage through static electrical discharge and have to be handled very cautiously with this in mind. Some arrays are imperfectly "passivated," i.e., they are liable to change their behavior if exposed to ordinary levels of atmospheric water vapor. The degradation produced in this way is, however, generally reversible by baking the array at moderate temperatures under high vacuum.

6.6 Efficiency of a system

The design of an infrared astronomical instrument is governed by the need to maximize the signal-to-noise ratio of its output. Clearly it is necessary to keep the throughput of photons as high as possible and to minimize the extraneous background.

Table 6.7. *Summary of efficiency considerations*

Atmospheric transmission
Reflectivity of mirrors
Telescope secondary mirror obscuration
Dewar window transmission
Transmission of lenses
Transmission of filter
Efficiency of diffraction grating
Detector QE

6.6.1 Throughput

In an instrument such as a camera, spectrograph or photometer, the throughput will be the product of factors listed in Table 6.7.

6.6.2 Choice of pixel size

The individual picture elements of an array are called *pixels*. For optimum accuracy and sensitivity the focal-plane scale of the telescope must be matched to the pixel size. For example, if the seeing is about 1 arcsec FWHM, the pixel scale should be about 2–3 pixels per arcsec to ensure good sampling for accurate photometry. On the other hand, if adaptive optics are in use, a much larger number of pixels per arcsec may be appropriate, since the images themselves will be smaller.

In general, it is undesirable to over-sample an image because the read-out noise will increase according to the number of pixels involved in each stellar image.

Many camera designs incorporate focal reducers of different ratios so that an appropriate choice can be made according to the prevailing seeing. As the numbers of pixels in an array increase, the demands on the optical designs of focal reducers become more severe. The wavelength range coverable by an InSb detector may require that two focal reducer lenses are used so that their designs can be optimized for particular parts of it.

6.6.3 The unwanted background

For ground-based instruments, the background against which the faint astronomical sources must be detected arises from the sources listed in

Table 6.8. *The unwanted background*

Atmospheric emissivity
Emission of telescope mirrors
Emission of telescope structure in beam
Emissivity of warm windows
Emissivity of surfaces within the cryogenic vessel
Scattered light within instrument

Table 6.8. The noise from these sources is equal to the square root of the number of photons emitted per second.

If the background seen by a detector arises entirely from objects at ambient temperature, then in the near-infrared a change of temperature can make a considerable difference to its level because of the rapid rise of the blackbody curve at its short-wavelength end. An increase in temperature from 273 K to 300 K will change the signal-to-noise ratio by 1.1 mag at K and 0.7 mag at L for faint sources.

Background measurements by Wainscoat and Cowie (1992) on the 2.2 m telescope of the University of Hawaii on Mauna Kea at an ambient temperature of 275 K yielded for the K' band 14.65 mag arcsec^{-2} from the telescope and instrument. The atmosphere at 273 K yielded 14.7 mag arcsec^{-2}, so that the total background was about $K' = 14.0$ arcsec^{-2}. At K, the total is about 13.4 mag arcsec^{-2} and at H it is about 14.0, with fluctuations of order of a factor of 2 in less than half an hour.

Low-temperature telescopes and detector surroundings remain essential if the full advantages of space-borne instrumentation are to be realized. It becomes possible to cool the telescope itself to very low temperatures without fear of condensation of atmospheric gases, and very low backgrounds may be obtained.

6.7 Seeing in the infrared
6.7.1 Atmospheric turbulence

The angular resolution of all but the smallest telescopes is limited by the turbulence of the Earth's atmosphere, which gives rise to the blurring of images called *seeing*. Ideally, the image of a point-like source in the focal plane of a telescope should be a classical diffraction pattern with its strong central peak. In practice, the time-averaged image of such a source resembles a two-dimensional Gaussian distribution. The usual measure of seeing quality is the full width at half-maximum (FWHM) of the distribution. The quality of the image produced by an optical system may also be described by its *Strehl ratio*, which is the ratio of its intensity on axis to that which would be produced if there were no aberrations.

Seeing in the infrared is somewhat better than that in the visible. Provided the intrinsic minimum given by the diffraction limit of the telescope does not come into account, it is found that

$$\text{image size} \propto \lambda^{-0.2}.$$

Thus at 16 μm the images should have half the extension that they have at 0.5 μm.

6.7.2 Seeing compensation

By means of *adaptive optics*, a means for compensating the aberrations introduced by atmospheric turbulence within the telescope, the seeing quality can be improved. The possibilities offered by this technique are much more easily realized at infrared wavelengths than in the visible.

The wavefronts of the light from a star as they enter a large telescope are bent and distorted from their passage through the Earth's atmosphere, so that as the light strikes the mirror it is in phase over only small regions of the pupil, causing the instantaneous image of a point source to appear as a series of "speckles." The scale length over which phase coherence is preserved is called the *Fried parameter*, r_0. It is a function of wavelength λ (and zenith angle ζ):

$$r_0(\lambda, \zeta) \propto \lambda^{6/5} f(\zeta).$$

At 0.5 μm the Fried parameter is typically 15 cm, while at the K-band under the same conditions (seeing $\sim$0.75 arcsec, etc.) it is 79 cm, according to an example given by Beckers (1993). In the near-infrared, the image formed by a 2 m or similar-sized telescope in good seeing is dominated by a single large speckle moving about in the focal plane.

6.7.3 Isoplanatic angle

An important concept in seeing compensation is the *isoplanatic angle*, namely the radius of a circle in the sky over which the atmospheric wavefront disturbances can be considered identical. It is thus a measure of the size of the region over which diffraction-limited images might be obtained with ideal adaptive optics. In the same example from Beckers (1993) given above, the isoplanatic angle is 1.9 arcsec at *V* and 10 arcsec at *K*. A further quantity, the *isokinetic angle*, refers to the angle over which the image has the same motion.

The timescale of image motion is proportional to the Fried parameter, so that the reaction time needed in the adaptive optics can be longer in the infrared than at shorter wavelengths.

6.7.4 Guide stars

The operation of adaptive optics depends on having a suitably bright guide star from which the corrections to the wavefront may be determined. Thus only fields within the isoplanatic angle of such guide stars can be observed in this way. Although the requirements ease with increasing wavelength, the usable regions are quite limited.

The use of artificial guide stars generated by lasers has been demonstrated. The "guide star" is generated in a layer of enhanced Na and K at a height of ~90 km, above the distortion-producing layers. Some account of military work in this field has been given by Fugate and Wild (1994). Lloyd-Hart et al. (1998) present some astronomical images which have been sharpened using this technique.

6.7.5 Tip-tilt and software seeing compensation

The lowest-order distortion produced by the atmosphere amounts to a linear shift of the image in the focal plane of the telescope. This term is a large fraction of the image degradation produced, and it is relatively easy to design equipment to compensate for it (Close and McCarthy, 1994). The isoplanatic field that applies to translational motion of the image is larger than that for full (diffraction-limited) correction. The response time of the compensation system may be longer and guide stars from a larger area of sky may be used. Close and McCarthy describe an adaptive tip-tilt secondary system and camera called FASTTRAC, attached to the 2.3 m telescope at Steward Observatory. They report a factor of 2 improvement in the Strehl ratio in the *H*-band, using a guide camera operating at *K*.

Tip-tilt compensation can also be applied in software when the read-out rate of an array is sufficiently rapid. The displacement of a bright star in the field of an individual exposure can be measured and used to counter-shift the digital data before co-adding.

6.8 Some representative instruments

In the following four sections some simple, representative, instruments are discussed. The infrared accessories for the 8–10 m telescopes completed or nearing completion (see Table 6.10 below) are similar in principle to those described, but differ in terms of size, versatility and engineering complexity.

6.8.1 Photometer

Simple single-channel photometers still offer the best method for obtaining high-precision photometry of moderately bright uncrowded objects (Fig. 6.5). Traditionally, star-sky chopping was accomplished by a rotating multi-bladed mirror and a fixed mirror in the focal plane, but this arrangement led to extra noise at longer wavelengths (from difraction off the moving mirror edges and imperfect matching of the pupil in the two positions) and defocussing of the image. Chopping is normally now accomplished by means of a stepwise-oscillating secondary mirror within the telescope, which effectively creates two positions on the sky, or beams, viewed alternately by the detector at a frequency of order 10 Hz. The output from the detector is passed to a phase-sensitive detector (lock-in amplifier) which automatically subtracts the signal in one sky patch from that in the other, synchronously with the chopping action. The difference is integrated and recorded. By moving the telescope periodically, say every 20 s, so that the object is placed first in one beam and then in the other, residual "second-order" background can be subtracted away. (This background arises from the fact that the detector "sees" slightly different parts of the telescope structure in the two positions of the chopping secondary.)

6.8.2 Camera

The essential parts of an infrared camera are illustrated by means of the simple instrument shown in Fig. 6.6. The field lens lies in the focal plane of the telescope and forms an image of the exit pupil on a cold stop, close to whichever re-imaging lens is in use. Re-imaging allows for magnifications of 1:1 and 1:0.5, to match different seeing conditions or fields of view. Baffles are employed to reduce the effect of stray light and the filter wheel is skewed to avoid spurious reflections. Cameras designed for larger arrays mainly differ by having more elaborate optical components. For example, fast doublet re-imaging lenses distort too

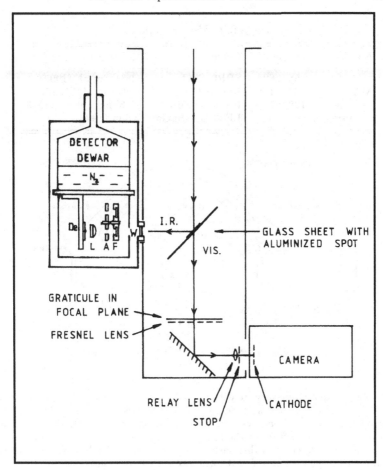

Fig. 6.5. Schematic layout of a SAAO photometer. A large field around the object of interest can be viewed by the CCD acquisition camera through a focal reducer consisting of the Fresnel and relay lenses. During acquisition of the object, the aluminized spot is moved off axis to view the central part of the field. While the measurement is progressing, the chopping secondary of the telescope causes two alternating images to be formed, one of which is seen by the detector and the other of which is viewed by the acquisition camera for guiding. Light enters the detector Dewar through a CaF_2 window (W) and passes through a filter (F), an aperture (A) and a Fabry or field lens (L) before reaching the InSb detector. The upper part of the Dewar contains N_2 slurry at about 60 K for cooling the optical parts.

Table 6.9. *Mid-IR cameras*

Name	Format	Detector	Reference
	58 × 62	SBRC Si:Ga	Gezari et al. (1992)
MANIAC	128 × 128	Rockwell Si:As	Böker et al. (1997)
MIRAC	128 × 128	Rockwell Si:As	Hoffmann et al. (1993)
TIMMI	64 × 64	LETI-LIR Si:Ga	Käufl (1993)

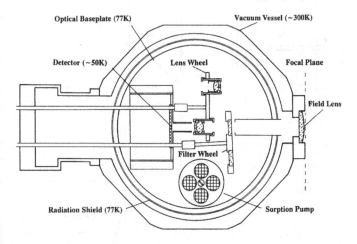

Fig. 6.6. Simple camera used with a 64 × 64 element array. There are two possible focal-plane scales given by interchangeable re-imaging lenses. To minimize distortion, the rays had to be kept paraxial, which required that the cold stop was in front of the re-imaging lens in one case and behind it in the other. The detector of this particular camera was a Philips HgCdTe array, with long-wavelength cutoff at 4.1 μm and a CCD readout.

much when used with larger arrays, and better focussing of the telescope exit pupil on the cold stop is achieved with an achromatic field lens.

Some descriptions of mid-infrared cameras are listed in Table 6.9. See also section 6.5.7 for a discussion of backgrounds and sensitivities in ground-based mid-infrared arrays.

6.8.2.1 Array camera sensitivity

The sensitivity of an infrared camera operating under background-limited conditions can be estimated from the following formula, based on Poissonian statistics, for the signal-to-noise ratio in combination with a known value on some particular telescope with the same thermal and night-sky background per arcsec^2:

$$S/N \propto D \left(\frac{t\eta}{N_{\text{arcsec}^2}} \right)^{1/2},$$

where D is the diameter of the telescope, t is the integration (exposure) time, η is the quantum efficiency of the detector and N_{arcsec^2} is the number of arcsec^2 over which the image is spread. The last factor, N_{arcsec^2}, enters because the level of the background is proportional to the area covered by the image.

As a reference point we can take the SOFI near-infrared camera attached to the 3.5 m NTT of ESO and equipped with a 1024×1024 Hawaii HgCdTe array. According to Dr. A. F. M. Moorwood (private communication), this instrument offers a sensitivity in $0.75''$ seeing, with an aperture of $1.5''$ diameter and 1 h integration time, of $J = 22.9$, $H = 21.9$ and $K_S = 20.9$ at a signal-to-noise of 5. The ambient temperature at the time of measurement was rather high ($16\,$C). An improvement of $\sim$0.5 mag might be expected at K when $T = 0\,$C.

6.8.3 Spectrometers

In a spectrometer, the area of the sky to be measured is defined by a cold slit or aperture. The next component in the optical path is a collimator which renders the rays parallel before they strike a diffraction grating at the image of the telescope exit pupil. The diffracted rays are then focussed by a "camera" onto the detector array.

There are many variations on this simple theme. A high-dispersion spectrograph can be constructed by having an echelle grating and an order-separating prism, so that a long spectrum can be placed in parallel segments onto a detector array (Fig. 6.7).

A grism (grating–prism combination in a single optical element) can be used in a camera to give spectroscopic coverage of a whole field. The prism component deflects the spectrum at mid-range to compensate for the deflection of the grating, so that the combination acts as an in-line disperser. Examples of such an instrument are the NICMOS camera of the Hubble Space Telescope (Axon et al., 1996) and the IRIS camera of the Anglo-Australian Observatory (Allen et al., 1993).

Efficient baffling is even more important in a spectrometer than in other infrared instruments, because each pixel of a camera is sensitive to the whole range of incoming background radiation while only receiving a very small fraction of the desired spectrum.

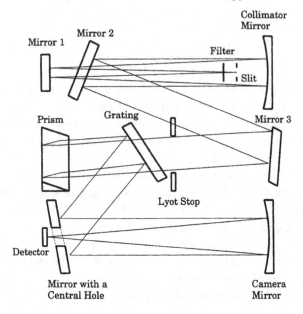

Fig. 6.7. Optical layout of a cooled echelle spectrometer. This is a two-dimensional view of a three-dimensional layout by Imanishi et al. (1996). The instrument uses four plane mirrors to fold the optical train. It operates from 2 to 5 μm. The detector is a 256 × 256 InSb array. Cooling is achieved with a mechanical refrigerator.

Astronomical spectrometers for the 8 m and other large telescopes have been described in the literature. Their optical layout is usually complicated by the need for versatility (many dispersions and wavelength ranges) and compactness, which involves "folding" the long beam paths with plane mirrors.

Ground-based spectrometers operating at wavelengths where the thermal background is high can be used for sources only moderately brighter than can be observed by cameras operating at similar wavelengths because the background seen by each spectral element is reduced in the same proportion as the signal. This is not, of course, true for situations when the background is low and readout noise is the limiting factor.

In practice, sensitivity can be calculated as for a camera but allowing for the inefficiency of the extra optical components, especially the grating. The entrance slit may also be undersized for good resolution, at the expense of the amount of light entering the spectrometer, and the response pattern at a particular wavelength may be spread over many pixels.

Table 6.10. *Large telescopes under construction or nearing completion*

Telescope	diameter (m)	Mirror type	Altitude (m)
Gemini N&S	8	ULE	4100, 2715
HET	≤9.2	91 segments Zd	2000
Keck I and II	9.8	36 segments Zd	4100
Large binocular	2 × 8.4	Borosilicate spun	3200
Magellan I and II	6.5	Borosilicate spun	2300
MMT upgrade	6.5	Borosilicate spun	2600
Subaru	8.3	ULE	4100
VLT	4 × 8.2	Zerodur (Zd)	2600

6.8.3.1 Fourier transform spectrometers

Fourier transform spectrometers, based on the Michelson principle, offer very high resolution but are confined to the examination of bright objects. They are uncompetetive for faint ones because they use one or two detectors which receive the photon noise of all wavelengths (within the range selected by cold pre-filters), whereas more modern spectrometers are based on arrays and each detector receives only the flux associated with its wavelength interval. The achievements of the Fourier transform spectrometer have been summarized by Ridgway and Brault (1984).

6.8.4 Large telescopes and instruments

Most of the new large telescopes (Table 6.10) have been planned to exploit the infrared region.

The Gemini instruments in Chile and Hawaii, typical of the new large telescopes, incorporate several features designed to enhance their infrared performance. These include protected silver coatings on the mirrors and effective cleaning systems to keep their emissivities low. Their secondaries will have tip-tilt actuators, and the quality of their optics will be such as to yield nearly diffraction-limited performance in the K-band.

A listing of some new instruments developed or being developed for large telescopes is given in Table 6.11.

Table 6.11. *Some newly developed instruments for large telescopes*

Instrument	Telescope	Function	Reference
ISAAC	ESO VLT	1–5 μm imaging, spectroscopy, polarimetry	Moorwood (1995)
CONICA	ESO VLT	1–5 μm camera, adaptive optics	Lenzen & Hofmann (1995)
CRIRES	ESO VLT	$R \sim 100000$ echelle spectrometer	Wiedemann et al. (1995)
NIRSPEC	Keck II	1–5 μm spectroscopy	McLean et al. (1995)
IRCS	Subaru	1–5 μm camera & spectrograph	Nishimura et al. (1997)
COMICS	Subaru	Mid-IR camera and spectrograph	Nishimura et al. (1997)

6.9 Observing and data reduction

Before planning an observing program, it is necessary to understand the procedures involved in data reduction.

In array camera work, a typical exposure contains, besides the desired image, background contributions from telescope and atmospheric radiation. In addition, the sensitivity (quantum efficiency) will vary across the array and there will be numbers of bad pixels, which may be dead (insensitive) or have excessive dark current (and hence noise).

The typical night might include flat field exposures for each band in use, telescope focus exposures, standard stars, program objects and comparison fields.

The well-known data reduction packages such as IRAF (Kitt Peak National Observatory), MIDAS (ESO), FIGARO (AAO) and STARLINK (Rutherford-Appleton Laboratory) contain routines which can be used for infrared image reduction. Among these is the important facility to mosaic a group of small-area images into a large final product.

Some on-line reduction facilities should be available for checking each image immediately after acquisition, in case a repeat exposure should be necessary. For example, it is possible to subtract a previous frame, taken with the same filter and exposure time, to get a rough image. A few stellar images can be examined rapidly by Gaussian fit to check the focus, the seeing and the image quality before moving on to the next exposure.

6.9.1 Array problems

The on-chip output amplifiers of many arrays have the undesirable property that they glow in the infrared, causing pixels in their neighborhood to show high background levels. This effect can be mitigated by operating the output circuitry at the lowest possible currrent levels for proper functioning and by switching it off during exposures.

Arrays frequently suffer from other problems such as memory effects, which cause images from previous exposures to re-appear as ghosts on later ones, especially when its the electron storage capacity has been saturated by exposure to excessively bright sources. This problem can sometimes be reduced or eliminated by multiple "pre-wipes" to read out the charge very thoroughly.

Long-wavelength arrays on satellites often change their characteristics due to bombardment with energetic charged particles, such as are encountered when passing through the van Allen Belts, and some form of annealing, such as a partial warming-up, may be necessary after each such occasion.

The behavior of the NICMOS camera and arrays on HST has been investigated in depth and several papers have been devoted to it in the 1997 HST Calibration Workshop (Casertano et al., 1997).

6.9.2 Linearizing

As mentioned in section 6.5, the data from infrared arrays may need to be linearized before further processing.

6.9.3 Removal of instrumental and sky background

In principle, for a well-behaved detector and atmosphere, a frame can be cleaned of extraneous background by subtracting a sky frame of equal exposure time, made at a nearby position that does not contain astronomical objects. For large-format detectors, suitable patches of blank sky may be impossible to find.

In reality, for ground-based observations, the sky background varies continuously, whether because of changes in temperature or OH airglow emission (see section 2.3.1). The latter can vary on timescales of a few minutes. Because of this, exposures are frequently short. It is often better to make dithered exposures followed by median averaging than to attempt to move the telescope to a blank piece of sky for background determination.

At long wavelengths, very short exposures may be made in synchronism with chopping. The individual exposures may be added to, or subtracted from, an image accumulator according to the position of the chopper. In this way the first-order background is reduced. The telescope may also be moved periodically, and a second subtraction performed, to eliminate residual effects caused by differences in the thermal radiation seen by the detector in the two positions of the chopper.

6.9.4 Dithering: dealing with bad pixels

Dithering refers to a technique where several exposures are made with the position of the telescope slightly displaced each time, relative to the field center. The displacements may be larger or smaller than the pixel spacing, according to the type of problem to be overcome.

6.9.4.1 Overcoming background variations

If dithering is used to overcome background variations, exposures may be made with shifts greater than the size of a stellar image, so that the median averaging process works properly. Of course, if the field is very crowded, chance coincidences of images on the shifted frames may lead to median averages which do not represent the true background.

6.9.4.2 Isolated bad pixels

When there are many isolated bad pixels, a set of exposures with the telescope displaced successively by one pixel width may be obtained. During data processing, the numerical images may be counter-displaced. A perfect final image is then formed by median averaging the re-centered images pixel by pixel. Isolated anomalous pixels are eliminated in this way. Alternatively, bad pixels may be mapped and "bridged over" by replacing them with the average value of surrounding good pixels, though this is not to be recommended.

6.9.4.3 Undersampling of images

When a stellar image is undersampled, i.e., there are too few pixels per image, it can be profitable to make multiple images with sub-pixel displacements of the field. This ensures better photometry by reducing effects due to non-uniform sensitivity across pixels or dead spaces between them.

6.9.5 Sky frame from median averaging

The *median average* of a set of measurements is the value which has equal quantities of individual measurements above and below it. For large samples with a Poissonian distribution, it approaches the mean value.

As mentioned, a sky frame without stars is often impossible to obtain. An "empty" sky frame may be generated by median averaging a number of frames of different fields, obtained with the same exposure time and filter, etc. The stellar images are eliminated by this process if the fields are reasonably sparsely populated. However, before median averaging can take place, it is usually necessary to adjust each pixel of the frame by a uniform amount for variations in the background level, so that the mode (most frequently obtained pixel value) of each is the same.

In survey work, determination of the sky level by median averaging saves time by removing the necessity of taking equal numbers of empty frames.

The subtraction of a sky frame will, of course, increase the random background noise of the reduced data. However, by taking a large number of background frames and median averaging them, this source of additional noise can be made negligible.

6.9.5.1 Flat fielding

The sensitivity of the detector can be normalized by dividing it by a flat field frame. Such a frame is constructed by observing a uniform source, for example, a screen or a featureless piece of sky, and subtracting a dark frame of equal duration. In the case of a screen, a flat field free of telescope background and scattered light can be obtained by subtracting an exposure with the illumination off from one with it on. This process can be repeated a sufficient number of times and the results averaged to make the flat field essentially noise free.

6.9.5.2 Standardization

For accurate work, measurements of standard stars should be undertaken with sufficient frequency. These stars should be as near as possible in zenith angle to the measured object to reduce errors due to imperfectly known extinction coefficients. Ideally, they should be of similar color to the objects being measured, though this is not practicable for many objects of interest in the infrared. On-line reduction is valuable in assessing the photometric quality of the night and hence the frequency with which standards must be observed.

6.9.5.3 Extraction of photometric information

The positions and intensities of stellar images may be extracted efficiently and automatically by using a program such as DoPHOT (Schechter, Mateo and Saha, 1993) or DAOPHOT (Stetson, 1987). These programs make optimum fits to the images to reduce the contribution of pixels with insignificant data to the noise. Calibration must be done separately, for example, by using "aperture photometry" on the images and comparing the results with the standard values. Corrections for extinction as a function of airmass must be applied when standardizing the photometry.

6.10 Further reading

Infrared astronomical instruments are described frequently in issues of PASP.

The *Rockwell Science Center*, the manufacturer of many popular astronomical arrays, has an informative web site at http://www.risc.rockwell.com.

Santa Barbara Research Center also has a web site at http://www.sbrc.com.

SPIE, the Society of Photo-Optical Instrumentation Engineers, conducts astronomical infrared and other conferences from time to time, dealing with detectors and instrumentation (see, e.g., Fowler, 1995).

Fowler, A. M., 1995. *Infrared Detectors and Instrumentation for Astronomy*, SPIE, **2475**. A similar conference was held in March 1998 and is the subject of *SPIE*, **3356**.

Low, F. and Rieke, G. H., 1974. *Instrumentation and Techniques of Infrared Astronomy*, Chap. 9 of *Methods of Experimental Physics*, Vol. 12. Academic Press, New York & London. Old, but still useful.

McLean, I. S., (ed.), 1994. *Infrared Astronomy with Arrays: the Next Generation*, proc. of a conf. held at UCLA. Kluwer, Dordrecht.

McLean, I. S., 1997. *Electronic Imaging in Astronomy*, Wiley-Praxis, Chichester. Readable, with much technical detail.

Rieke, G. H., 1994. *Detection of Light: from the Ultraviolet to the Submillimeter*, Cambridge University Press. Physics of infrared and other detectors.

Wolf, W. L., Zissis, G. J., 1978. *The Infrared Handbook*, Office of Naval Research, Dept of the Navy, Washington, DC. A useful compendium of information for infrared system designers. Military orientated.

References

Aannestad, P. A. and Kenyon, S. J., 1979. *Astrophys. Space Sci.*, **65**, 155.

Aaronson, M., Frogel, J. A. and Persson, S. E., 1978. *Astrophys. J.*, **220**, 442.

Ali, B., Carr, J. S., DePoy, D. L., Frogel, J. A. and Sellgren, K., 1995. *Astron. J.*, **110**, 2415.

Allamandola, L. J., Sandford, S. A., Tielens, A. G. G. M. and Herbst, T. M., 1992. *Astrophys. J.*, **399**, 134.

Allard, F. and Hauschildt, P. H., 1995. *Astrophys. J.*, **445**, 433.

Allard, F., Hauschildt, P. H., Alexander, D. R. and Starrfield, S., 1997. *Ann. Rev. Astron. Astrophys.*, **35**, 137.

Allard, F., Hauschildt, P. H., Miller, S. and Tennyson, J., 1994. *Astrophys. J.*, **426**, L39.

Allen, C. A., 1963. *Astrophysical Quantities*, Athlone Press, London.

Allen, D. A. and Cragg, T. A., 1983. *MNRAS*, **203**, 777.

Allen, D. A. et al., 1993. *Proc. Astron. Soc. Australia*, **10**, 298.

André, P., 1994. In *The Cold Universe*, Montmerle, T., Lada, C. J., Mirabel, I. F. and Trân Tranh Vân, J. (eds.), Editions Frontières, Gif-sur-Yvette, p. 179.

André, P., Ward-Thompson, D. and Barsony, M., 1993. *Astrophys. J.*, **406**, 122.

Angione, R. J., 1989. In *Infrared Extinction and Standardization*, Milone, E. F. (ed.), Springer, Berlin, p. 25.

Ashley, M. C. B. et al., 1996. *PASP*, **108**, 721.

Axon, D., Calzetti, D., MacKenty, C. and Skinner, C., 1996. *NICMOS Instrument Handbook Ver. 1.0*, Space Tel. Sci. Inst., Baltimore.

Barlow, M. J., 1989. *Proc 22nd ESLAB Symposium on Infrared Spectroscopy in Astronomy*, Kaldeich, B. H. (ed.), ESA SP-290, p. 307.

Barlow, M. J. et al., 1996. *Astron. Astrophys.*, **315**, L241.

Barvainis, R., 1987. *Astrophys. J.*, **320**, 537.

Beckers, J. M., 1993. *Ann. Rev. Astron. Astrophys.*, **31**, 13.

Becklin, E. E., 1997. In *The Far Infrared and Submillimetre Universe*, Pillbratt, G., Volonte, S. and Wilson, A. (eds.), ESA SP-401, p. 201.

Beichman, C. A., Neugebauer, G., Habing, H. J., Clegg, P. E. and Chester, T. J., 1988. *Infrared Astronomical Satellite (IRAS), Catalogs and Atlases, Explanatory Supplement*, NASA RP–1190, Vol. 1.

Beintema, D. A. et al., 1996. *Astron. Astrophys.*, **315**, L369.

Bell, R. A. and Briley, M. M., 1991. *Astron. J.*, **102**, 763.

Bell, R. A. and Gustafsson, B., 1989. *MNRAS*, **236**, 653.

Bell, R. A., Gustafsson, B., Nordh, H. L. and Olofsson, S. G., 1976. *Astron. Astrophys.*, **46**, 391.

Bell, R. A. and Tripicco, M. J., 1991. *Astron. J.*, **102**, 777.

Bell Burnell, S. J., Davies, J. K. and Stobie, R. S. (eds.), 1992. *Astrophys. Sp. Sci. Revs.*, **61**, Nos. 1 & 2.

Bennett, J. M. and Ashley, E. J., 1965. *Applied Optics*, **4**, 221.

Bersanelli, M., Bouchet, P. and Falomo, R., 1991. *Astron. Astrophys.*, **252**, 854.

Bessell, M. S. and Brett, J. M., 1988. *PASP*, **100**, 1134.

Bessell, M. S., Brett, J. M., Scholz, M. and Wood, P. R., 1989a. *Astron. Astrophys. Suppl.*, **77**, 1.

Bessell, M. S., Brett, J. M., Scholz, M. and Wood, P. R., 1989b. *Astron. Astrophys.*, **213**, 209.

Bessell, M. S., Castelli, F. and Plez, B., 1998. *Astron. Astrophys.*, **333**, 231 (er. **337**, 321).

Blackwell, D. E., Leggett, S. K., Petford, A. D., Mountain, C. M. and Selby, M. J., 1983. *MNRAS*, **205**, 897.

Blackwell, D. E. and Shallis, M. J., 1977. *MNRAS*, **180**, 177.

Block, D. L., 1996. In Block, D. L. and Greenberg, J. M. (eds.), *New Extragalactic Perspectives in the New South Africa*, Kluwer Academic Publishers, Dordrecht, p. 1.

Bohlin, R. C., Savage, B. D. and Drake, J. F., 1978. *Astrophys. J.*, **224**, 132.

Böker, T., Storey, J. W. V., Krabbe, A. and Lehmann, T., 1997. *PASP*, **109**, 827.

Boogert, A. C. A. et al., 1996. *Astron. Astrophys.*, **315**, L377.

Borysov, A., Jørgensen, U. G. and Zheng, C., 1998. *Astron. Astrophys.*, in press.

Bouchet, P., Manfroid, J. and Schmider, F. X., 1991, *Astron. Astrophys. Suppl.*, **91**, 409.

Brocklehurst, M., 1971. *MNRAS*, **153**, 471.

Burrows, A. et al., 1997. *Astrophs. J.*, **491**, 856.

Burton, M. G. et al. (eds.), 1994. *Proc. Astron. Soc. Australia*, **11**, 127.

Burton, M. G., Hollenbach, D. J. and Tielens, A. G. G. M., 1992. *Astrophys. J.*, **399**, 563.

Campins, H., Rieke, G. H. and Lebofsky, M. J., 1985. *Astron. J.*, **90**, 896.

Carter, B. S., 1990. *MNRAS*, **242**, 1.

Carter, B. S. and Meadows, V. S., 1995. *MNRAS*, **276**, 734.

Casali, M. and Hawarden, T., 1992. *JCMT-UKIRT Newsletter*, No. 4, p. 33.

Casertano, S., Jedrzejewski, R., Keyes, T. and Stevens, M., 1997. *The 1997 HST Calibration Workshop*, Space Telescope Science Institute, Baltimore.

Cernicharo, J. et al., 1996. *Astron. Astrophys*, **315**, L201.

Chernoff, D. F. and McKee, C. F., 1990. In *Molecular Astrophysics: A Volume Honoring Alexander Dalgarno*, Hartquist, T. W. (ed.), Cambridge, p. 360.

Chini, R. and Krügel, E., 1996. In Block, D. L. and Greenberg, J. M. (eds.), *New Extragalactic Perspectives in the New South Africa*, Kluwer Academic Publishers, Dordrecht, p. 329.

Clavel, J., Wamsteker, W. and Glass, I. S., 1989. *Astrophys. J.*, **337**, 236.

Close, L. M. and McCarthy, D. W., Jr., 1994. *PASP*, **106**, 77.

Cohen, M., Walker, R. G., Barlow, M. J. and Deacon, J. R., 1992a. *Astron. J.*, **104**, 1650.

Cohen, M., Walker, R. G. and Witteborn, F. C., 1992b. *Astron. J.*, **104**, 2030.

Cohen, M., Witteborn, F. C., Walker, R. G., Bregman, J. D. and Wooden, D. H., 1995. *Astron. J.*, **110**, 275.

Cohen, M., Witteborn, F. C., Bregman, J. D., Wooden, D. H., Salama, A. and Metcalfe, L., 1996a. *Astron. J.*, **112**, 241.

Cohen, M., Witteborn, F. C., Carbon, D. F., Davies, J. K., Wooden, D. H. and Bregman, J. D., 1996b. *Astron. J.*, **112**, 2274.

Cohen, M. and Davies, J. K., 1995. *MNRAS*, **276**, 715.

Connolly, A. J., Csabai, I., Szalay, A. S., Koo, D. C., Kron, R. G. and Munn, J. A., 1995. *Astron. J.*, **110**, 2655.

Connolly, A. J., Szalay, A. S., Dickinson, M., SubraRao, M. U. and Brunner, R. J., 1997. *Astrophys. J.*, **486**, L11.

Dachs, J., Engels, D. and Kiehling, R., 1988. *Astron. Astrophys.*, **194**, 167.

D'Antona, F. and Mazzitelli, I., 1994. *Astrophys. J. Suppl.*, **90**, 467

de Graauw, Th. et al., 1996. *Astron. Astrophys.*, **315**, L345.

Devereux, N., 1996. In Block, D. L. and Greenberg, J. M. (eds.), *New Extragalactic Perspectives in the New South Africa*, Kluwer Academic Publishers, Dordrecht, p. 357.

Di Benedetto, G. P., 1993. *Astron. Astrophys.*, **270**, 315.

Draine, B. T., 1989. *Proc 22nd ESLAB Symp. on IR Spectroscopy in Astron.*, Kaldeich, B. H. (ed.), ESA SP-290, p. 93.

Draine, B. T. and Bertholdi, F., 1996. *Astrophys. J.*, **468**, 269.

Draine, B. T. and Lee, H. M., 1984. *Astrophys. J.*, **285**, 89.

Duley, W. W. and Williams, D. A., 1988. *MNRAS*. **231**, 969.

Dwek, E. (ed.), 1995. *Unveiling the Cosmic Infrared Background*, AIP Conf. Proc. 348, Amer. Inst. Phys., New York.

Edelson, R. A. and Malkan, M. A., 1986. *Astrophys. J.*, **308**, 59.

Ehrenfreund, P. et al., 1996. *Astron. Astrophys*, **315**, L341.

Elias, J. H., Frogel, J. A., Matthews, K. and Neugebauer, G., 1982. *Astron. J.*, **87**, 1029 (err. 1893).

Elias, J. H., Frogel, J. A., Hyland, A. R. and Jones, T. J., 1983. *Astron. J.*, **88**, 1027.

Ellis, R. S., 1997. *Ann. Rev. Astron. Astrophys.*, **35**, 389.

Engels, D., Sherwood, W. A., Wamsteker, W. and Schultz, G. V., 1981. *Astron. Astrophys. Suppl.*, **45**, 5.

Epchtein, N. et al. 1994. *Astrophys. Sp. Sci.*, **217**, 3.

Feast, M. W., Whitelock, P. A. and Carter, B. S., 1990. *MNRAS*, **247**, 227.

Ferguson, J. W., Korista, K. T. and Ferland, G. J., 1997. *Astrophys. J. Suppl.*, **110**, 287.

Ferland, G. J., 1980. *PASP*, **92**, 596.

Ferland, G. J., 1996. *Hazy, a Brief Introduction to Cloudy*, University of Kentucky, Dept. of Physics Internal Report. See also http://www.pa.uky.edu/~gary/cloudy.

Feuchtgruber, H. et al., 1997. *Astrophys. J.*, **487**, 962.

Finger, G., Nicolini, G. P., Meyer, M. and Moorwood, A. F. M., 1995. In Fowler, A. M. (ed.), *Infrared Detectors and Instrumentation for Astronomy*, SPIE, **2475**, 15.

Fink, U. and Larson, H. P., 1979. *Astrophys. J.*, **233**, 1021.

Fowler, A. M. and Gatley, I., 1990. *Astrophys. J.*, **353**, L33.

Fowler, A. M., Gatley, I., Vbra, F. J., Ables, H. D., Hoffman, A. and Woolaway, J., 1994. In McLean, I. (ed.), *Infrared Astronomy with Arrays*, Kluwer, Dordrecht, p. 399.

Fowler, A. M., Heynssens, J. B., Gatley, I., Vbra, F. J., Ables, H. D., Hoffman, A. W. and Woolaway, J. T., 1995. In Fowler, A. M. (ed.), *Infrared Detectors and Instrumentation for Astronomy*, SPIE, **2475**, 27.

Frogel, J. A., Persson, S. E., Aaronson, M. and Matthews, K., 1978. *Astrophys. J.*, **220**, 75.

Frogel, J. A., Terndrup, D. M., Blanco, V. M. and Whitford, A. E., 1990. *Astrophys. J.*, **353**, 494.

Frogel, J. A. and Whitford, A. E., 1987. *Astrophys. J.*, **320**, 199.

Fugate, R. Q. and Wild, W. J., 1994. *Sky and Telescope*, **87**, No. 5, 25.

Garnett, J. D. and Forrest, W. J., 1993. In Fowler, A. M. (ed.), *Infrared Detectors and Instrumentation*, SPIE, **1946**, 395.

Geballe, T. R., 1990. In *Molecular Astrophysics: A Volume Honoring Alexander Dalgarno*, Hartquist, T. W. (ed.), Cambridge, p. 345.

Geballe, T. R., 1997. In *From Stardust to Planetesimals*, Pendleton, Y. J. and Tielens, A. G. G. M. (eds.), Astron. Soc. Pacific Conf. Ser. **122**, 119.

Geballe, T. R., Kulkarni, S. R., Woodward, C. E. and Sloan, G. C., 1996. *Astrophys. J.*, **467**, L101.

Genzel, R. et al., 1998. *Astrophys J.*, **498**, 579.

Gezari, D. Y., Folz, W. C., Woods, L. A. and Varosi, F., 1992. *PASP*, **104**, 191.

Gezari, D. Y., Schmitz, M., Pitts, P. S. and Mead, J. M., 1993a. *Catalog of Infrared Observations 3rd Ed.*, NASA Ref. Publ. 1294.

Gezari, D. Y., Schmitz, M., Pitts, P. S. and Mead, J. M., 1993b. *Far Infrared Supplement, Catalog of Infrared Observations 3rd Ed.*, NASA Ref. Publ. 1295, Rev. 1.

Giles, K., 1977. *MNRAS*, **180**, 57p.

Glass, I. S., 1974. *Mon. Notes Astron. Soc. Sthn. Africa*, **33**, 53 (err. 71).

Glass, I. S., 1984. *MNRAS*, **211**, 461.

Glass, I. S., 1985. *Irish Astron. J.*, **17**, 1.

Glass, I. S., 1997. *Mon. Notes Astron. Soc. Sthn. Africa*, **56**, 110.

Glass, I. S. and Carter, B. S., 1989. in *Infrared Extinction and Standardization*, Milone, E. F. (ed.), Springer, Berlin, p. 37.

Glass, I. S., Catchpole, R. M. and Whitelock, P. A., 1987. *MNRAS*, **227**, 373.

Glass, I. S. and Moorwood, A. F. M., 1985. *MNRAS*, **214**, 429.

Glass, I. S., Sekiguchi, K. and Nakada, Y., 1995. *IAU Symp. 167, New Developments in Array Technology and Applications*, Philip, A. G. D., James, K. A. and Upgren, A. R. (eds.), Kluwer, Dordrecht, p. 109.

Golay, M., 1974. *Introduction to Astronomical Photometry*, D. Reidel Publishing Co., Dordrecht, the Netherlands.

Gray, D. F., 1976. *The Observation and Analysis of Stellar Photospheres*, Wiley, New York.

Greenberg, J. M. and Li, A., 1996. In Block, D. L. and Greenberg, J. M. (eds.), *New Extragalactic Perspectives in the New South Africa*, Kluwer Academic Publishers, Dordrecht, p. 118.

Greene, T. P. and Lada, C. J., 1996. *Astron. J.*, **112**, 2184.

Greenhouse, M. A. et al., 1991. *Astrophys. J.*, **383**, 164.

Groenewegen, M. A. T., Whitelock. P. A., Smith, C. H. and Kerschbaum, F., 1998. *MNRAS*, **293**, 18.

Habing, H. J., 1996. *Astron. Astrophys. Revs.*, **7**, 97.

Habing, H. J. et al., 1996. *Astron. Astrophys*, **315**, L233.

Hamann, F., DePoy, D. L., Johansson, S. and Elias, J., 1994. *Astrophys. J.*, **422**, 626.

Hamann, F. and Persson, S. E., 1989. *Astrophys. J. Suppl.*, **71**, 931.

Hanel, R. A. et al., 1986. *Science*, **233**, 70.

Hanson, M. M., Conti, P. S., Rieke, M. J., 1996. *Astrophys. J. Suppl.*, **107**, 281.

Hayes, K. M., LaViolette, R. A., Stapelbroeck, M. G. and Petroff, M. D., 1989. In *Proc. 3rd IR Detector Technology Workshop*, McCreight, C. (ed.), NASA Tech. Memo. 102209.

Henning, Th., Michel, B., Stognienko, R., 1995. *Planetary and Space Science*, **43**, 1333.

Herbst, T. M., 1994. *PASP*, **106**, 1298.

Herschel, Sir W., 1800. *Phil. Trans. Roy. Soc.*, **90**, 284.

Herter, T., 1989. *Proc 22nd ESLAB Symposium on Infrared Spectroscopy in Astronomy*, Kaldeich, B. H. (ed.), ESA SP-290, p. 403.

Herzberg, G., 1971. *The Spectra and Structures of Simple Free Radicals, an Intro. to Molecular Spectroscopy*, Cornell University Press, Ithaca, N.Y., Repr. Dover Publs. Inc., New York, 1988.

Hildebrand, R. H. and Dragovan, M., 1995. *Astrophys. J.*, **450**, 663.

Hinkle, K., Wallace, L. and Livingston, W., 1995. *PASP*, **107**, 1042.

Hodapp, K.-W. et al., 1995. In *Infrared Detectors and Instrumentation for Astronomy*, Fowler, A. M. (ed.), SPIE **2475**, 8.

Hoessel, J. G., Elias, J. H., Wade, R. A. and Huchra, J. P., 1979. *PASP*, **91**, 41.

Hoffmann, W. F. et al., 1993. *SPIE*, **1946**, 334.

Hollenbach, D. J. and Tielens, A. G. G. M., 1997. *Ann. Rev. Astron. Astrophys.*, **35**, 179.

Hough, J. H., Peacock, T. and Bailey, J. A., 1991. *MNRAS*, **248**, 74.

Hough, J. H., Crysostomou, A. and Bailey, J. A., 1994. *Experimental Astronomy*, **3**, 127.

Hummer, D. G. and Storey, P. J., 1987. *MNRAS*, **224**, 801.

Hunt, L. K. et al., 1998. *Astron. J.*, **115**, 2594.

Imanishi, M. et al., 1996. *PASP*, **108**, 1129.

Israel, F. P. (ed.), 1986. *Light on Dark Matter, Proc. 1st IRAS Conf., held in Noordwijk, the Netherlands, 10-14 June 1985*, Reidel, Dordrecht.

Jennings, R. E., 1975. In *HII Regions and Related Topics*, Lecture Notes in Physics No. 42, Wilson, T. L. and Downes, D., (eds.), Springer, Berlin, p. 137.

Johnson, H. L., 1962. *Astrophys. J.*, **135**, 69.

Johnson, H. L., 1964. *Bol. Ton. y Tac.*, **3**, 305.

Johnson, H. L., 1965. *Comm. Lunar and Planet. Lab.*, **3**, 73.

Johnson, H. L., 1966. *Ann. Rev. Astron. Astrophys.*, **4**, 193.

Johnson, H. L., MacArthur, J. W. and Mitchell, R. I., 1968. *Astrophys. J.*, **152**, 465.

Johnson, H. L., Mitchell, R. I., Iriarte, B. and Wiśniewski, W. Z., 1966. *Comm. Lunar & Planetary Lab.*, No. 63, **4**, 99.

Jones, T. J. and Hyland, A. R., 1982. *MNRAS*, **200**, 509.

Jones, H. R. A., Longmore, A. J., Jameson, R. F. and Mountain, C. M., 1994. *MNRAS*, **267**, 413.

Jørgensen, U. G., 1997. In *Molecules in Astrophysics*, IAU Symp. 178, van Dishoeck, E. F. (ed.), p. 441.

Justtanont, K., Skinner, C. J. and Tielens, A. G. G. M., 1994. *Astrophys. J.*, **435**, 852.

Justtanont, K. and Tielens, A. G. G. M., 1992. *Astrophys. J.*, **389**, 400.

Käufl, H. U., 1993. *ESO Messenger*, No. 72, p. 42.

Kleinmann, S. G. et al., 1994. *Astrophys. Space Sci.*, **217**, 11.

Kleinmann, S. G. and Hall, D. N. B., 1986. *Astrophys. J. Supp.*, **62**, 501.

Koorneef, J., 1983a. *Astron. Astrophys. Suppl.*, **51**, 489.

Koorneef, J., 1983b. *Astron. Astrophys.*, **128**, 84.

Kozlowski, L. J., 1996. In *Infrared Readout Electronics III, Proc. SPIE*, **2745**, 1.

Kwok, S., 1993. *Ann. Rev. Astron. Astrophys.*, **31**, 63.

Lada, C. J., 1987. In *Star Forming Regions, IAU Symp. 115*, Peimbert, M. and Jugaku, J. (eds.), Kluwer, Dordrecht, p. 1.

Lambert, D. L., Gustafsson, B., Eriksson, K. and Hinkle, K. H., 1986. *Astrophys. J. Suppl.*, **62**, 373.

Lançon, A. and Rocca-Volmerange, R., 1992. *Astron. Astrophys. Suppl.*, **96**, 593.

Landini, M., Natta, A., Oliva, E., Salinari, P. and Moorwood, A. F. M., 1984. *Astron. Astrophys.*, **134**, 284.

Laney, C. D. and Stobie, R. S., 1993. *MNRAS*, **263**, 921.

Lee, T. A., 1970. *Astrophys. J.*, **162**, 217.

Leggett, S. K., 1992. *Astrophys. J. Suppl.*, **82**, 351.

Leinert, Ch. et al., 1998. *Astron. Astrophys. Suppl.*, **127**, 1.

Leitherer, C. et al., 1996. *PASP*, **108**, 996.

Lenzen, R. and Hofmann, R., 1995. In *Infrared Detectors and Instrumentation for Astronomy*, Fowler, A. M. (ed.), SPIE, **2475**, 268.

Lenzuni, P., Chernoff, D. F. and Salpeter, E. E., 1991. *Astrophys. J. Suppl.*, **76**, 759.

Lloyd-Hart, M. et al., 1998. *Astrophys. J.*, **493**, 950.

Lopez, B. et al., 1997. *Astrophys. J.*, **488**, 807.

Lord, S., 1992. *Tech. Memo. 103957*, NASA.

Low, F. J., 1961. *J. Opt. Soc. Am.*, **51**, 1300.

Low, F. J. and Rieke, G. H., 1974. In *Methods of Experimental Physics*, Carleton, N. (ed.), Academic Press, New York, **12A**, 415.

Lucas, C. et al., 1995. *Proc. SPIE*, **2475**, 50.

Luginbuhl, C. B., Henden, A. A., Vrba, F. J. and Guetter, H. H., 1995. In Fowler, A. (ed.), *Infrared Detectors and Instrumentation for Astronomy*, *Proc. SPIE*, **2475**, 88.

Ludwig, C. B., 1971. *Applied Optics*, **10**, 1057.

Luhman, K. L. and Rieke, G. H., 1996. *Astrophys. J.*, **461**, 298.

Lutz, D. et al., 1996. *Astron. Astrophys.*, **315**, L269.

Lutz, D., Krabbe, A. and Genzel, R., 1993. *Astrophys. J.*, **418**, 244.

Macleod, H. M., 1986. *Thin-film optical filters*, 2nd ed., Adam Hilger, Bristol (now Institute of Physics Publishing).

Madau, P., Ferguson, H. C., Dickinson, M. E., Giavalisco, M., Steidel, C. C. and Fruchter, A., 1996. *MNRAS*, **283**, 1388.

Maihara, T., Iwamuro, F., Yamashita, F., Hall, D. N. B., Cowie, L. L., Tokunaga, A. T. and Pickles, A., 1993. *PASP*, **105**, 940.

Maiolino, R., Rieke, G. H. and Rieke, M. J., 1996. *Astron. J.*, **111**, 537.

Maloney, P. and Black, J. H., 1988. *Astrophys. J.*, **325**, 389.

Manduca, A. and Bell, R. A., 1979. *PASP*, **91**, 848.

Mann, R. G. et al., 1997. *MNRAS*, **289**, 482.

Marconi, A., van der Werf, P. P., Moorwood, A. F. M. and Oliva, E., 1996. *Astron. Astrophys.*, **315**, 335.

Martin P. G. and Whittet, D. C. B., 1990. *Astrophys. J.*, **357**, 113.

Mathis, J. S., Rumpl, W. and Nordsieck, K. H., 1977. *Astrophys. J.*, **217**, 425.

Matsumoto, T., Matsuura, S. and Noda, M., 1994. *PASP*, **106**, 1217.

Mattila, K., et al. 1996. *Astron. Astrophys.*, **315**, L353.

McGregor, P. J., 1994. *PASP*, **106**, 508.

McGregor, P. J., Hyland, A. R. and Hillier, D. J., 1988. *Astrophys. J.*, **324**, 1071.

McLean, I. S., Becklin, E. E., Figer, D. F., Larson, S. and Liu, T., 1995. In *Infrared Detectors and Instrumentation for Astronomy*, Fowler, A. M. (ed.), SPIE, **2475**, 350.

Mégessier, C., 1994. *Astron. Astrophys.*, **289**, 202.

Mezger, P. G. and Henderson, A. P., 1967. *Astrophys. J.*, **147**, 471.

Mobasher, B., Sharples, R. M. and Ellis, R. S., 1993. *MNRAS*, **263**, 560.

Moorwood, A. F. M., 1995. In *Infrared Detectors and Instrumentation for Astronomy*, Fowler, A. M. (ed.), SPIE, **2475**, 262.

Moorwood, A. F. M., 1996. *Space Science Revs.*, **77**, 303.

Moorwood, A. F. M., 1997. In *Extragalactic Astronomy in the Infrared*, Mamon, G. A., Thuan, Trinh Xuan and Trân Tranh Vân, J. (eds.), Editions Frontières, Gif-sur-Yvette, p. 301.

Moorwood, A. F. M. and Salinari, P., 1983. *Astron. Astrophys.*, **125**, 342.

Morris, P. W., Eenens, P. R. J., Hanson, M. M., Conti, P. S. and Blum, R. D., 1996. *Astrophys. J.*, **470**, 597.

Mould, J. R. and Hyland, A. R., 1976. *Astrophys. J.*, **208**, 399.

Mullen, D. J., Pomerantz, M. A. and Stanev, T. (eds.), 1989. *Astrophysics in Antarctica* AIP Conference Proc., 198, Amer. Inst. Phys., New York.

Neufeld, D. A. et al., 1996. *Astron. Astrophys.*, **315**, L237.

Neugebauer, G. and Leighton, R. B., 1969. *Two-Micron Sky Survey – A Preliminary Catalog*, NASA SP-3047, Washington, D.C.

Neugebauer, G. et al., 1984. *Astrophys. J.*, **278**, L1.

Nguyen, H. T. et al., 1996. *PASP*, **108**, 718.

Nishimura, T. et al., 1997. In *Optical Telescopes of Today and Tomorrow*, Arneberg, A. (ed.), SPIE **2871**, 1064.

Offner, A., 1975. *Optical Engineering*, **14**, 130.

Okuda, H., Matsumoto, T. and Roellig, T. L., 1997. *Diffuse Infrared Radiation and the IRTS; ASP Conf. Ser.*, Vol. 124.

Oliva, E. and Gennari, S., 1995. *Astron. Astrophys. Suppl.*, **114**, 179.

Oliva, E. and Gennari, S., 1998. *Astron. Astrophys. Suppl.*, **128**, 599.

Oliva, E. and Moorwood, A. F. M., 1990. *Astrophys. J.*, **348**, L5.

Oliva, E. and Origlia, L., 1992. *Astron. Astrophys.*, **254**, 466.

Olnon, F. M., Raimond E. et al., 1986. *Astron. Astrophys. Suppl.*, **65**, 607.

Omont, A., Petitjean, P., Guilloteau, S., McMahon, R. G., Solomon, P. M. and Pécontal, E., 1996. *Nature*, **382**, 428.

176   *References*

Origlia, L., Moorwood, A. F. M. and Oliva, E., 1993. *Astron. Astrophys.*, **280**, 536.
Osterbrock, D. E., 1989. *Astrophysics of Gaseous Nebulae and Active Galactic Nuclei*, University Science Books, Mill Valley, California.
Oudmaijer, R. D., Waters, L. B. F. M., van der Veen, W. E. C. J. and Geballe, T. R., 1995. *Astron. Astrophys.*, **299**, 69.
Panagia, N., 1973. *Astron. J.*, **78**, 929.
Persson, S. E., Aaronson, M. and Frogel, J. A., 1977. *Astron. J.*, **82**, 729.
Persson, S. E., Frogel, J. A. and Aaronson, M., 1979. *Astrophys. J. Suppl.*, **39**, 61.
Persson, S. E., Murphy, D. C., Krzeminski, W., Roth, M. and Rieke, M. J., 1998. *Astron. J.*, **116**, 2475.
Persson, S. E., West, S. C., Carr, D. M., Sivaramakrishnan, A. and Murphy, D. C., 1992. *PASP*, **104**, 204.
Poggianti, B. M., 1997. *Astron. Astrophys. Suppl.*, **122**, 399.
Pottasch, S., 1993. In *Infrared Astronomy*, Mampaso, A., Prieto, M. and Sánchez, F. (eds.), Cambridge University Press, Cambridge, p. 63.
Pradhan, A. K. and Zhang, H. L., 1993. *Astrophys. J.*, **409**, L77.
Price, S. D., 1977. *The AFGL Four-Colour IR Sky Survey: Supplementary Catalog.* AFGL-TR-77-0160, AS A067 017.
Price, S. D. and Murdock, T. L., 1983. *The Revised AFGL Infrared Sky Survey Catalog, AFGL-TR-83-0161*, Air Force Geophysics Lab., Hanscomb Field, Mass, USA.
Price, S. D., 1988. *PASP*, **100**, 171.
Price, S. D. and Walker, 1976. *The AFGL Four-Colour IR Sky Survey: Catalog of Observations at 4.2, 11.0, 19.8 and 27.4 μm.* AFGL-TR-76-0208.
Puget, J. L. and Léger, A., 1989. *Ann. Rev. Astron. Astrophys.*, **27**, 161.
Ramírez, S. V., DePoy, D. L., Frogel, J. A., Sellgren, K. and Blum, R. D., 1997. *Astron. J.*, **113**, 1411.
Ramsay, S. K., Mountain, C. M. and Geballe, T. R., 1992. *MNRAS*, **259**, 751.
Ridgway, S. T. and Brault, J. W., 1984. *Ann. Rev. Astron. Astrophys.*, **22**, 291.
Ridgway, S. T., Carbon, D. F. and Hall, D. N. B., 1978. *Astrophys. J.*, **225**, 138.
Ridgway, S. T., Joyce, R. R., White, N. M. and Wing, R. F., 1980. *Astrophys. J.*, **235**, 126.
Rieke, G. H., 1994. *Detection of Light: from the Ultraviolet to the Sub-millimeter*, Cambridge University Press.
Rieke, G. H. and Lebofsky, M. J., 1985. *Astrophys. J.*, **288**, 618.
Rieke, G. H., Lebofsky M. J. and Low, F. J., 1985. *Astron. J.*, **90**, 900.
Rinsland, C. P. and Wing, R. F., 1982. *Astrophys. J.*, **262**, 201.
Roche, P. F. and Aitken, D. K., 1985. *MNRAS*, **215**, 425.
Roelfsema, P. R. et al., 1996. *Astron. Astrophys.*, **315**, L289.
Rowan-Robinson, M. et al., 1997. *MNRAS*, **289**, 490.
Rudy, R. J., Erwin, P., Rossano, G. S. and Puetter, R. C., 1991. *Astrophys. J.*, **383**, 344.
Salpeter, E. E., 1977. *Ann. Rev. Astron. Astrophys.*, **15**, 267.
Sanders, D. B. and Mirabel, I. F., 1996. *Ann. Rev. Astron. Astrophys.*, **34**, 749.

Sanders, D. B., Phinney, E.S., Neugebauer, G., Soifer, B. T. and Matthews, K., 1989. *Astrophys. J.*, **347**, 29; erratum 1990, **357**, 291.

Sandford, S. A., Allamandola, L. J. and Bernstein, M. P., 1997. In *From Stardust to Planetesimals*, Pendleton, Y. J. and Tielens, A. G. G. M. (eds.), Astron. Soc. Pacific Conf. Ser. **122**, 201.

Schaeidt, S. G. et al., 1996. *Astron. Astrophys.*, **315**, L55.

Schechter, P. L., Mateo, M. and Saha, A., 1993. *PASP*, **105**, 1342.

Scholz, M., 1985. *Astron. Astrophys.*, **145**, 251.

Schönberner, D., 1983. *Astrophys. J.*, **272**, 708.

Schutte, W. A. and Tielens, A. G. G. M., 1989. *Astrophys. J.*, **343**, 369.

Selby, M. J., Mountain, C. M., Blackwell, D. E., Petford, A. D. and Leggett, S. K., 1983. *MNRAS*, **203**, 795.

Sellgren, K., 1984. *Astrophys. J.*, **277**, 623.

Shibai, H. and Murakami, H., 1996. In *Infrared Technology and Applications XII*, Andresen, B. F. and Scholl, M. S. (eds.), SPIE **2744**, 75.

Simon, M. and Cassar, L., 1984. *Astrophys. J.*, **283**, 179.

Simon, T., Morrison, D. and Cruikshank, D. P., 1972. *Astrophys. J.*, **177**, L17.

Simons, D., 1996. *Near Infrared Filter Bandpasses for Gemini Instruments*, Kitt Peak Natl. Obs., TN-PS-G0037.

Simpson, J. P., 1975. *Astron. Astrophys.*, **39**, 43.

Smith, C. H., Aitken, D. A. and Moore, T. J. T., 1994. in *Instrumentation in Astronomy VIII*, Crawford, D. R. and Craine, E. R. (eds.), SPIE, **2198**, 736.

Sodrowski, T. J. et al., 1997. *Astrophys J.*, **480**, 173.

Spinoglio, L. and Malkan, M. A., 1992. *Astrophys. J.*, **399**, 504.

Stacey, G. J., 1989. In *Proc. 22nd ESLAB Symp. on Infrared Spectroscopy in Astron.*, Kaldeich, B. H. (ed.), ESA SP-290, ESA, Paris, p. 455.

Stapelbroeck, M. G., Seib, D. H., Huffman, J. E., Florence, R. A., 1995. *Proc. SPIE*, **2475**, 41.

Sterken, C. and Manfroid, J., 1992. *Astronomical Photometry, a Guide*, Kluwer, Dordrecht.

Sternberg, A., 1990. In *Molecular Astrophysics*, Hartquist, T. W. (ed.), Cambridge University Press, Cambridge, p. 384.

Stetson, P. B., 1987. *PASP*, **99**, 191.

Stewart, H. S. and Hopfield, R. F., 1965. In Kingslake, R. (ed.), *Applied Optics and Optical Engineering*, Academic Press, New York, **1**, 127.

Strecker, D. W., Erickson, E. F. and Witteborn, F. C., 1979. *Astrophys. J. Suppl.*, **41**, 501.

Struck-Marcell, C. and Tinsley, B. M., 1978. *Astrophys. J.*, **221**, 562.

Sutherland, R. S., Allen, M. G., Kewley, and Dopita, M. A, 1999. *Astrophys. J. Suppl.*, in press.

Swinyard, B. M. et al., 1996. *Astron. Astrophys.*, **315**, L43.

Tanaka, W., Hashimoto, O., Nakada, Y., Onaka, T., Tanabe, T., Okada, T., and Yamashita, Y., 1990. *Publ. Natl. Astron. Obsy. Japan*, **1**, 259.

Terndrup, D. M., Davies, R. L., Frogel, J. A., DePoy, D. L. and Wells, L. A., 1994. *Astrophys. J.*, **432**, 518; erratum 1995, **454**, 945.

Thomas, J. A., Hyland, A. R. and Robinson, G., 1973. *MNRAS*, **165**, 201.

Thompson, R. I., 1995. *Astrophys. J.*, **445**, 700.

Thuan, T. X., 1983. *Astrophys. J.*, **268**, 667.

Tielens, A. G. G. M., Waters, L. B. F. M, Molster, F. J. and Justtanont, K., 1998. *Astrophys. Space Sci.*, **255**, 415.

Timmermann, R. et al., 1996. *Astron. Astrophys.*, **315**, L281.

Tinbergen, J., 1996. *Astronomical Polarimetry*, Cambridge University Press, Cambridge.

Tinney, C. G., Mould, J. R. and Reid, I. N., 1993. *Astron. J.*, **105**, 1045.

Tokunaga, A. T., 1984. *Astron. J.*, **89**, 172.

Tokunaga, A. T., 1999. In *Astrophysical Quantities*, 4th edition, Cox, A. (ed.), Springer, submitted.

Traub, W. A. and Stier, M. T., 1976. *Appl. Optics*, **15**, 364.

Treuenfels, E. W., 1963. *J. Opt. Soc. Am.*, **53**, 1162.

Tsuji, T., Ohnaka, K. and Aoki, W., 1996. *Astron. Astrophys.*, **305**, L1.

Van de Hulst, H. C., 1946. *Rech. Astron. Obs. Utrecht*, **11**, 1.

Van de Hulst, H. C., 1957. *Light Scattering by Small Particles*, Wiley, New York.

Van der Veen, W. E. C. J. and Habing, H. J., 1988. *Astron. Astrophys.*, **194**, 125.

Van der Veen, W. E. C. J. and Olofsson, H., 1990. In *From Miras to Planetary Nebulae: Which Path for Stellar Evolution*, Menessier, M. O. and Omont, A. (eds.), Editions Frontières, Gif-sur-Yvette, p. 139.

Van Dishoeck, E. F. et al., 1996. *Astron. Astrophys.*, **315**, L349.

Venzon, J. E., Lum, N. A., Freeman, S. D. and Domingo, G., 1995. *Proc. SPIE*, **2475**, 34.

Volk, K., Clark, T. A. and Milone, E. F., 1989. In *Infrared Extinction and Standardization*, Milone, E. F. (ed.), Springer, Berlin, p. 15.

Volk, K. and Cohen, M., 1989. *Astron. J.*, **98**, 931.

Volk, K., Kwok, S., Stencel, R. E. and Brugel, E., 1991. *Astrophys. J. Suppl.*, **77**, 607.

Wainscoat, R. J. and Cowie, L. L., 1992. *Astron. J.*, **103**, 332.

Walker, R. G., 1969. *Phil. Trans. Roy. Soc.*, **A264**, 209.

Walker, R. G. and Price, S. D., 1975. *The AFCRL Sky Survey Catalog of Observations at 4, 11, and 20 μm*, AFCRL-TR-0375.

Wallace, L. and Hinkle, K., 1996. *Astrophys. J. Suppl.*, **107**, 312.

Wallace, L. and Hinkle, K., 1997. *Astrophys. J. Suppl.*, **111**, 445.

Wallace, L., Livingston, W., Hinkle, K. and Bernath, P., 1996. *Astrophys. J. Suppl.*, **106**, 165.

Wamsteker, W., 1981. *Astron. Astrophys.*, **97**, 329.

Wesselius, P. R. et al., 1996. *Astron. Astrophys.*, **315**, L197.

Whittet, D. C. B., 1987. *QJRAS*, **28**, 303.

Whittet, D. C. B., 1992. *Dust in the Galactic Environment*, IOP Publishing, Bristol.

Whittet, D. C. B. and Duley, W. W., 1991. *Astron. Astrophys. Rev.*, **2**, 167.

Whittet, D. C. B. and van Breda, I. G., 1978. *Astron. Astrophys.*, **66**, 57.

Whittet, D. C. B. and van Breda, I. G., 1980. *MNRAS*, **192**, 467.

Whittet, D. C. B., Bode, M. F. and Murdin, P., 1987. *Vistas in Astronomy*, **30**, 135.

Whittet, D. C. B. et al., 1996. *Astron. Astrophys.*, **315**, L357.

Wiedemann, G., Delabre, B. and Moorwood, A. F. M., 1995. In *Infrared Detectors and Instrumentation for Astronomy*, Fowler, A. M. (ed.), SPIE, **2475**, 279.

Wolf, W. L. and Zissis, G. J., 1978. *The Infrared Handbook*, Office of Naval Research, Dept. of the Navy, Washington, D.C.

Wood, D. O. S. and Churchwell, E., 1989. *Astrophys. J. Suppl.*, **69**, 831.

Yamamura, I. et al., 1998. *Astrophys. Sp. Sci.*, **255**, 351.

Yen, V. L., 1970, *Infrared Handbook*, published by OCLI (Optical Coating Laboratory Inc., PO Box 7397, Santa Rosa, CA 95407).

Young, A. T., Milone, E. F. and Stagg, C. R., 1994. *Astron. Astrophys. Suppl.*, **105**, 259.

Young, J. S. and Scoville, N. Z., 1982. *Astrophys. J.*, **258**, 467.

Index

Printed in the United States
By Bookmasters